Safety Critical Systems

IEEE Press
445 Hoes Lane
Piscataway, NJ 08854

Safety Critical Systems

A Practical Guide

Mario L. Papini

Rebus Solutions S.R.L.
Firenze, Italy

IEEE Press
WILEY

Published by John Wiley & Sons, Inc., Hoboken, New Jersey.

Library of Congress Cataloging-in-Publication Data has been applied for:

Hardback ISBN: 9781394232062
ePDF ISBN: 9781394232086
ePub ISBN: 9781394232079
oBook ISBN: 9781394232093

Cover Design: Wiley
Cover Image: akindo/Getty Images

To my family—thanks.

Contents

About the Author

Mario L. Papini has spent over 35 years working on rail signaling and safety-critical systems. *Safety Critical Systems: A Practical Guide* is what that experience looks like written down.

He started at Alcatel Transport Automation in 1989 as a programmer, eventually becoming the technical go-between for the project management team and the customer. Then came stints at Lea+Elliott and SYSTRA in the United States, preparing signaling specs for airport people-mover contracts—detailed, unglamorous, consequential work.

He joined Harmon Industries, which was later acquired by GE Transportation in 2000. Over the course of 13 years, he held several positions: project manager, regional manager, EMEA Engineering Manager with R&D responsibility across Europe, the Middle East, and Africa, then Implementation Director for a team that built a new hardware/software platform from scratch.

In 2014, he co-founded Rebus Solutions S.R.L. along with his wife, Evelyn Steiner, and lives near Florence. The company provides embedded computer vision and consulting for safety-critical systems. He still runs it.

He has a B.Eng. in electrical engineering and a B.Sc. in computer science from the University of Saskatchewan. He works in English and Italian.

AI Declaration

The author acknowledges that AI tools were used in the preparation of this manuscript in the following capacities:

- to generate the first draft of the illustrations (subsequently changed by hand by the illustrator) and
- to review the manuscript for consistency and to suggest edits for clarity.

AI tools were not used to generate any text that was directly incorporated into the book.

Acknowledgments

I would like to acknowledge and thank the many professionals that I have encountered and continue to encounter throughout my career, who took the time to explain and discuss things with a little engineer that had no idea what he was talking about. Without their unearned patience, this book would not have been possible. Special thanks to Evelyn Papini-Steiner for thoughtful review and feedback on early drafts.

Mario L. Papini

Acronyms

AI	Artificial intelligence
ALARP	As low as reasonably practical
CCB	Change control board
ECC	Engineering change control
FFT	Functional fault tree
FMEA	Failure mode and effects analysis
FMECA	Failure modes and effects criticality analysis
FTA	Fault tree analysis
LLM	Large language model
MISRA	Motor Industry Software Reliability Association
MTBHE	Mean time between hazardous events
NOK	Not OK
PHA	Preliminary hazard analysis
USD	United States dollars
V&V	Verification and validation

Introduction

Who this book is for. Practicing engineers, early-career safety practitioners, and students who need a practical, decision-focused companion—not a theory text or a math-heavy reference. The aim is to help you make sound safety calls in day-to-day work, with just enough theory to justify decisions.

What we mean by safety-critical systems. These are systems whose failure can cause severe harm (up to death) to people or major damage to assets or the environment. Chapter 1 formalizes this with severity and likelihood, and sets up the risk matrix used throughout.

How to read this book. You can read straight through or dip in by topic:

- Chapter 1 defines safety, risk, and the safety-critical context.
- Chapter 2 clarifies independence and why it matters for decisions and evidence.
- Chapters 3 and 4 cover change control and building a safety-critical system end-to-end.
- Chapters 5 and 6 dive into verification/validation and analytical methods, with worked examples.
- Chapters 7–9 address fallacies, pitfalls, and management practices.
- Chapter 10 focuses on people and organizational dynamics (including insider risk).
- Chapter 11 covers monitoring, maintenance, and operational support.
- Chapter 12 surveys emerging technologies (automotive, medical, aerospace, ICS, AI/LLMs) with domain standards.
- Chapter 13 wraps up with next steps and further resources.

Tone and expectations. The book favors illustrative examples and checklists to move quickly from concept to action. Examples are illustrative, not full case studies. Where data or standards are cited, in-text references point to full IEEE-style entries so you can trace the source. Standards change; check the latest revision for your domain.

Safety is a multidisciplinary field that spans industries and requires a tailored approach for each context. Many standards can be applied, but the basic process of safety sometimes gets lost in the standards, and people forget what they are doing, choosing instead to "follow the recipe" that is presented in the standards. The standards attempt to capture the collective experience of a specific industry but are not an excuse for putting one's brain in neutral and blindly following the standard. The Safety Process must be a structured sequence of critical thinking that builds our confidence in the system that we are producing. Decisions must be challenged and justified with informed opinions.

This book focuses on practical strategies for managing safety-critical systems, including:

- Understanding and applying safety principles
- Implementing robust design, verification, and validation processes (with references to real failures)
- Avoiding common fallacies and missteps in safety analysis
- Managing the challenges of safety-critical projects and organizations
- Addressing emerging technologies, such as self-driving cars and AI-based systems

Safety is not a product that can be added to a system; it is a process woven into every stage of ideation, design, development, operation, and maintenance of the system. By the end of this book, you will have the tools to navigate this process with confidence.

Part I

Foundations

1

It Has to Be Safe

After reading this chapter, you should be able to:

- Explain safety using context, hazard, and risk.
- Set a project risk matrix: severity bands, a clear time base, and use As Low As Reasonably Practical (ALARP).
- Classify a few example hazards and show your reasoning.
- Note any residual risk and who needs to know.

Whether it is a familiar requirement or a new challenge that has just arrived on your desk, making critical systems safe is difficult. It requires clear thinking and often adherence to national and international standards to mitigate both risks and liability. Throughout this book, a **safety-critical system** means one whose failure or malfunction can lead to death or severe injury, substantial environmental damage, or major asset loss [1]. When in doubt, treat the system as safety-critical until you have shown otherwise.

This book introduces the design and management of safety-critical systems. It explains the reasoning behind established practices rather than prescribing checklists. We will cover structured analysis and clear communication that work across industries. In examining the safety design process, we start by defining some key concepts.

1.1 What Is Safety?

Safety is the state of a system, within a defined context, that minimizes the likelihood of harm to people or significant damage to material. Safety is inherently probabilistic: absolute safety is unattainable, but we can reduce risks to tolerable levels (defined in the risk matrix below). See Figure 1.1 for a compact visual of this relationship. Safety-critical systems deserve special attention because the potential

Figure 1.1 Safety as a function of context.

severity of harm includes fatalities or life-altering injuries; we explicitly account for severity in the risk matrix below.

For simplicity, throughout this book, we can consider hazards to be "bad things" and risks as a combination of the likelihood and severity of these bad things.

For example, a spinning blade inside a closed blender is relatively safe; the same blade spinning with the top off is less so. The context in which a system operates is critical to defining its safety. For example, a knife is neither safe nor unsafe: it depends on how and where it is being used.

When discussing safety, establish a shared vocabulary with your collaborators. Every company and project has its own terminology. Misunderstandings can lead to significant errors, and errors can lead to "bad things." This book defines terms as they are introduced, using straightforward language wherever possible without referencing external standards. The terminology here is decoupled from standards jargon, so beginners can follow.

Example 1.1: *Consider a project developing a new chemical processing system. During a design review meeting, an engineer notes that a certain valve must "fail-safe." The software team implements this to mean the valve should remain open if control is lost, assuming this maintains process flow. However, the safety engineer meant the valve should close upon failure to prevent chemical release. This terminology misalignment could lead to a hazardous condition where chemicals continue to flow during a system failure. A precise definition stating "fail-safe for valve X means the valve automatically closes upon loss of control signal or power" would have prevented this misunderstanding.*

Few things are more confusing than sitting through a long meeting attempting to resolve an issue, only to find that the two parties were using the same word to mean two different things. In the railway domain, I have seen it often when referring to a "train": do you mean the entire train, an individual carriage, the locomotor(s) or something else? As innocent as this might seem, it can lead to dramatic situations in which the same words mean very different things to each party.

If you are starting a new project, ensure that all participants (stakeholders) align on the definitions that the project will use from the outset. A system or project glossary of terms should be standard in any safety process. Often, if the project is governed by a contract, the contract will contain the base definitions that are applicable to the project. When contract definitions conflict with internal ones, communicate the difference to everyone in the organization to minimize confusion.

1.2 The Limits of Safety

A safety-critical system cannot be safe under all conditions. As we have previously noted, a number of factors, both internal and external, can impact the safety of a system. Safety is constrained by factors such as:

- Changes in context or environment: A toaster that works perfectly well on a countertop for its intended purpose could introduce a shock hazard in a wet environment.
- Poor maintenance practices (erosion or degradation of the system's performance).
- Misuse of the system: Such as using industrial equipment without proper training or removing safety guards from machinery.
- Multiple independent component failures: The mechanisms that maintain safety are subject to failures that should be monitored and mitigated. Simple failure chains are considered in the design but complex failure chains are not.
- Flawed design or implementation: This doesn't refer to just "bugs" in software or errors in design, but also to overly complex design choices that are beyond our current ability to analyze completely.

Safety is not absolute. It is impacted by the environment in which the system operates, so you must clearly define the boundaries of a system's safety. The safety of a system must be quantified and bound by its intended context and operating constraints. Without clear boundaries, risks to users, or even bystanders, may go unnoticed until it is too late.

Your system operates in the context of other systems; identifying these interfaces and making them explicit helps you find what could affect safety.

Example 1.2: *Consider an automatic door system designed for an office building. The system is designed to detect approaching individuals and open before they reach the door. The safety boundaries include:*

- *Environmental conditions: The sensors are rated for indoor use between 0 and 40 °C with normal lighting conditions.*
- *User expectations: Users will approach at walking speed from the front.*

- *Maintenance: Monthly cleaning of sensors and quarterly inspection of mechanical components.*

Outside these boundaries—such as in heavy rain affecting the sensors, users running at the door from an angle, or after months of neglected maintenance—the safety of the system cannot be guaranteed. Explicitly documenting these boundaries helps users understand when the system might not behave as expected.

1.3 Quantifying Safety

1.3.1 Risk Matrix

1.3.1.1 Hazard Severity (Impact)

Use the first applicable class.

I. Catastrophic: Loss of life or life-altering injury to ≥1 person.
II. Major: Serious injury requiring extended recovery or major equipment damage (≥US$100,000).
III. Minor: Injury requiring limited recovery (≤1 week) or modest equipment damage (<US$100,000).
IV. Incidental: First-aid only; no time off work.

1.3.1.2 Hazard Probability (Likelihood)

Pick *one time base* for the project and state it at the top of the table. Recommended options: *per operating hour, per mission,* or *per year*. Then define the classes with that base:

A. Frequent: Expected to occur >1× per T.
B. Occasional: Expected to occur ≤1× per T.
C. Remote: Expected to occur ≤1× per $5T$.
D. Improbable: Not expected within the defined operating life (state the life, e.g., 20 years), but credible.

Here T is the chosen time base (e.g., one mission, one year).

1.3.1.3 Risk Matrix and Decision Language

The matrix below preserves the I–IV × A–D logic and makes the ALARP expectation explicit. Place the chosen time base in the caption. Use these bands as a starting point; adjust thresholds per domain standards (e.g., ISO 26262, IEC 61508) and state the source.

1.3.1.4 Decision Rules

- **Unacceptable:** Reduce risk—change design, controls, or scope.
- **Tolerable (ALARP):** Permitted only with documented best-practicable risk reduction and evidence; record residual risk and justification.
- **Acceptable:** No further reduction required; maintain controls and monitor.

1.3.1.5 Caption and Footnotes

Likelihood classes are defined per the stated time base; the system's operating life must be specified (e.g., 20 years). ALARP evidence may include test data, analysis, inspection, and operational controls; choose the appropriate mix and document it.

To manage safety effectively, you need to quantify it. Quantifying safety shows you the expected level of safety at which the system will operate and provides a basis for evaluating your design choices. Quantification of safety is typically done using risk analysis, which evaluates two aspects:

- Hazard Severity: How bad is the "bad thing?"
- Hazard Probability: How often could the bad thing happen?

Example 1.3: ***Mini-example.*** *Suppose the hazard is "loss of braking assistance." If it is judged Severity II (serious injury possible) and Probability class C (unlikely but credible), then, using Table 1.1, the combination falls in the region typically treated as "tolerable if ALARP." In practice, that means either reducing probability (e.g., fault detection and graceful degradation) or reducing severity (e.g., limiting achievable hazardous energy).*

These two elements combine to establish a category summarizing the expected impact on people and things. Quantification of safety matters because safety must be considered as part of an engineering process, and engineering often requires

Table 1.1 Risk acceptance matrix (likelihood defined per stated time base).

Severity/ Probability	A	B	C	D
I	**Unacceptable**	**Unacceptable**	**Unacceptable**	Tolerable (ALARP)
II	**Unacceptable**	**Unacceptable**	Tolerable (ALARP)	Acceptable
III	**Unacceptable**	Tolerable (ALARP)	Acceptable	Acceptable
IV	Tolerable (ALARP)	Acceptable	Acceptable	Acceptable

trade-offs. Without a quantification of safety, how do we know if we are getting better or doing worse?

For practical purposes, risks are categorized into levels of severity and probability. Consider the following example of hazard severity:

I. Catastrophic: Loss of life or life-changing injuries to one or more individuals.
II. Major: Injuries requiring up to two months of recovery or significant equipment damage exceeding US$100,000.
III. Minor: Injuries requiring up to a week of recovery or equipment damage below US$100,000.
IV. Incidental: Minor injuries requiring no time off of work.

Similarly, hazard probability may be divided into:

A. Likely to occur more than once per month.
B. Likely to occur at most once per year.
C. Likely to occur at most once every five years.
D. Not expected to occur during the operating life of the system (20 years).

The combination of severity and probability yields a risk matrix that guides decision-making by categorizing risks as acceptable, tolerable, or unacceptable. At the beginning of a project, you must establish this criteria and determine for your industry, sector or application what is or is not acceptable. This should be formalized in the project as a risk assessment table.

The example risk assessment table (Table 1.2) shows the level of risk that the project is willing to undertake. The categorization as to whether a risk is acceptable or not can be quantified as OK or NOK (Not OK), but I have found that some flexibility is useful. The table above makes use of **"Unacceptable"** which means that you must fix this, changing the design or project goals if necessary, **"Tolerable"** which means that we can live with it the way it is, but there isn't a lot of margin and **"Acceptable"** which indicates that this is OK the way that it is. The objective of the overall project must be that at the end of the project, there are no more

Table 1.2 Example risk assessment table.

Severity/ probability	A	B	C	D
I	**Unacceptable**	**Unacceptable**	**Unacceptable**	Tolerable
II	**Unacceptable**	**Unacceptable**	Tolerable	Acceptable
III	**Unacceptable**	Tolerable	Acceptable	Acceptable
IV	Tolerable	Acceptable	Acceptable	Acceptable

"Unacceptable" risks present. Values shown are illustrative; use the applicable sector standard and state your chosen thresholds and sources in project documentation.

Key Takeaways

- Safety is argued with evidence, not assumed; absolute safety is unattainable—reduce risk to tolerable levels.
- Context defines safety: distinguish hazards (what can go wrong) from risk (likelihood × severity) and anchor terms with examples.
- Decisions depend on time base and operating life: set a project risk matrix and apply ALARP with recorded rationale.
- Shared vocabulary prevents unsafe ambiguity: define terms explicitly and keep terminology consistent across the team.
- Safety is a structured process of critical thinking: challenging decisions and justifying them with informed judgment.

2

Independence

After reading this chapter you should be able to:

- Point out independence types (process, analytical, organizational, etc.) in a real setup.
- Check independence using a short list: authority, conflicts, skills, reporting, and traceability.
- Lay out simple roles for V&V and change decisions with clear hand-offs.

Independence is a fundamental concept in the design, verification, and management of safety-critical systems. Independence ensures that processes, analyses, and system components operate free from undue influence or interdependence. Properly understood and applied, independence enhances the robustness of designs, simplifies validation, and optimizes resource allocation. The goal is better decisions and clearer evidence—not isolation for its own sake.

Independence is often misunderstood or misused. It can be invoked to justify shortcuts, assert process robustness, or simplify calculations without adequate demonstration. To ensure integrity, independence must be rigorously defined, challenged, and proven.

2.1 Defining Independence

At its core, independence can be described as "freedom of X from Y." For clarity, it is necessary to define:

1. What is independent (X): The entities or processes being evaluated.
2. From what (Y): The source of potential dependence.
3. How (type of freedom): The quantifiable way in which independence is established.

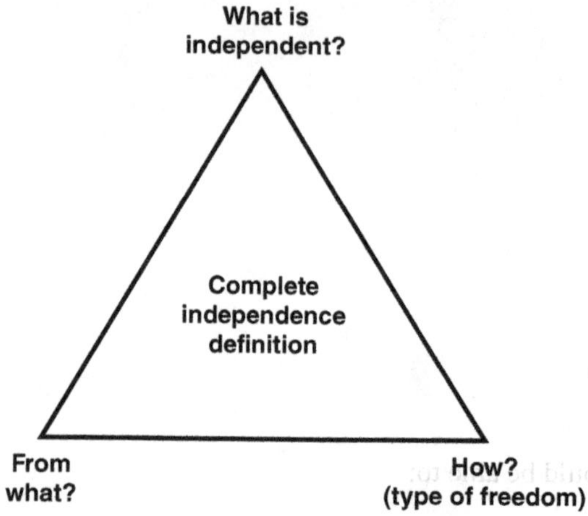

Figure 2.1 The three dimensions of independence.

See Figure 2.1 for a compact view of the dimensions discussed here.

Independence is not commutative.

For example, "Alice is financially independent from Bob" specifies the entities (Alice and Bob), and the type of dependence (financial freedom).

Example 2.1: *Consider a railway signaling system where two different computer systems (X and Y) are used to determine if a section of track is safe for train passage. Their independence could be defined as:*

- *What is independent: Computer system X*
- *From what: Computer system Y*
- *Type of freedom: Hardware independence (separate processors, power supplies, and I/O systems), software independence (different algorithms developed by different teams), and failure independence (no common failure modes)*

This precise definition allows engineers to verify the specific aspects of independence and avoid ambiguity in safety assessments.

Understanding independence is critical as it allows us to understand what we have really gained and what we might be missing. Faulty assumptions about independence have led to significant failures in safety-critical systems. For example, assuming that failures in subsystems are independent without rigorous analysis can result in cascading failures.

2.2 Types of Independence

Independence exists in several forms, each vital to safety-critical systems. Although there may be other types of independence possible, the following sections describe the primary types of independence that should be considered when discussing safety-critical systems.

2.2.1 Process Independence

Process independence ensures that different steps of the design life-cycle are carried out without undue influence on one another. This is typically achieved by having the two activities make use of the same input information and then compare their outputs only when they are complete. An example is the separation between design and testing. Typically, both activities begin with a set of requirements as input and when both activities are complete, they then influence each other (the test procedures are applied to the design output). The two interpretations of the requirements (the test branch and the design branch) are compared during testing. If there is a disagreement between the design and test, either design is wrong, test is wrong or the original requirement was wrong (see Figure 2.2).

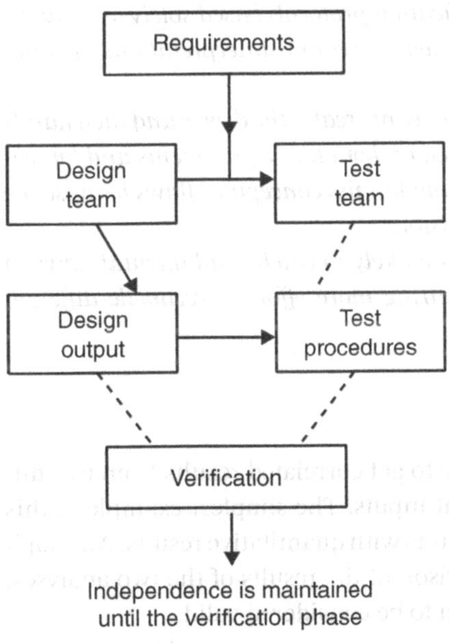

Figure 2.2 Process independence in design and testing.

- Independence in a test process brings diverse perspectives to a process that includes an implicit or explicit interpretation step, leveraging different experiences and reducing blind spots. For instance:
  - In integration testing, both designers and testers start from the same requirements but apply different interpretations, enhancing robustness.
  - In unit testing,[1] the tester validates implementation details, with lower independence as the focus is on verifying specific outputs.
- Project Management Implications: High process independence allows parallel execution, reducing schedule risks. However, it increases the potential for divergent interpretations, which must be reconciled during testing.

If the design and test were not independent, only one perspective on the requirements would be used if the same group and skill-set were used, or worse, if the designer generated the tests. In cases where there is low process independence, we run the risk of believing that risk has been reduced, while in fact, because there was no independence, we are still carrying the risk that we had hoped to mitigate.

Example 2.2: *A medical device manufacturer develops a new insulin delivery system. Two approaches to process independence could be:*

Strong Process Independence: *Requirements are developed and then handed to both the design and test teams who work independently. The design team creates the device, while the test team develops verification protocols based solely on requirements. When testing occurs, discrepancies might reveal misinterpretations in either the design or the testing approach.*

Weak Process Independence: *The design team creates the device and then hands it to the testing team, who develops tests based on both the requirements and knowledge of the implementation. This approach might miss conceptual flaws because the test development is influenced by design decisions.*

The strong independence approach is more likely to catch fundamental flaws in the requirements interpretation, while requiring more effort to reconcile different interpretations.

2.2.2 Analysis Independence

Analysis independence refers to the ability to get correlated results from two different analysis methods that have different inputs. The simplest example of this concept is the comparison of qualitative results with quantitative results. Although this will not yield precision in the comparison of the results of the two analyses, the results must correlate in order for them to be considered valid.

---

1 Unit testing being a form of software white-box testing in which detailed implementation is checked, including branch and decision coverage.

Example 2.3: *Consider a braking system for a train. Two different analyses are performed:*

Qualitative Analysis: *A failure modes and effects analysis (FMEA) identifies critical failure modes and ranks them by severity, likelihood, and detectability. This analysis concludes that brake actuator failures and signal transmission errors are the highest risks.*

Quantitative Analysis: *A probabilistic risk assessment using fault tree analysis calculates the probability of brake failure at 1×10^{-7} per hour, with signal transmission contributing 60% of this risk and actuator failures contributing 25%. These analyses show independence because they use different methodologies and data sources, yet they correlate in identifying the same primary risk factors. If they had identified entirely different risks as critical, this would suggest a problem in one or both analyses.*

- Low Independence Example: Failure modes and effects criticality analysis (FMECA) relies heavily on design outputs, limiting its robustness. Despite this, FMECA is essential for tracing failures across multiple design levels.
- High Independence Example: Preliminary hazard analysis (PHA) starts from system requirements, ensuring independence from specific design decisions. It is most effective when involving domain experts from diverse fields.

2.2.3 Organizational Independence

Organizational independence ensures that teams operate free from another team's influence. This is typically achieved through independent reporting structures. The concern that this is attempting to address is that if the teams work together too much, then they begin to see things in the same way, thus undermining the other forms of independence that we are discussing. If abused, organizational independence becomes the basis for isolating teams. Isolation and independence are two very different concepts.

- Challenges: True independence can be difficult to maintain, even with external teams, for example, when auditors work repeatedly with the same organizations, the level of independence is eroded.
- Practical Benefits: Even marginal organizational independence prevents undue pressure and ensures that verification findings are addressed transparently.
- Project Management Implications: Project managers have to accept a minimum level of inefficiency that comes about due to repeated functions in different teams. For example, each independent team may need separate

document control systems, configuration management procedures, and quality checks, creating redundancy but ensuring no single point of failure in the quality assurance process.

Organizational independence is not seeking to ensure that teams don't talk. It is trying to ensure that they don't start to think in the same way and, instead of having constructive tension between the teams that they subtly and inadvertently pull in the same direction during decision making.

Example 2.4: *A pharmaceutical company develops a critical manufacturing process for a new drug. They establish organizational independence through:*

Development Team: *Reports to the R&D Director, focuses on creating an efficient, high-yield manufacturing process.*

Quality Assurance Team: *Reports to the Quality Director (who reports directly to the CEO), and is responsible for verifying that the process meets all safety and regulatory requirements.*

This organizational structure ensures that quality concerns cannot be overridden by development pressures. During a production milestone review, the QA team identified potential contamination risks that weren't addressed in the design. Despite schedule pressure, the QA team can maintain their position because they don't report to the same management chain that is responsible for meeting development timelines.

2.2.4 System Independence

System independence refers to the degree of interaction between subsystems. Systems with low interaction are more independent, reducing the risk of cascading failures.

- Demonstration: Independence is typically proven through checklists and analyses of potential interactions. For example:
  - Shared power supplies indicate dependence.
  - Shared susceptibility to electromagnetic interference requires specific mitigations.
  - Shared components create common failure modes.

Example 2.5: *Consider an aircraft with redundant flight control computers. True system independence would require:*

Hardware Independence: *Separate processors, memory, and I/O systems with no common components.*

Software Independence: *Different software implementations, possibly using different algorithms to achieve the same function.*

Physical Independence: *Physical separation to prevent a single event (fire, impact, etc.) from affecting both systems.*

Power Independence: *Separate power sources with no common failure modes.*

Environmental Independence: *Resistance to common environmental factors like temperature, vibration, or electromagnetic interference.*

Without addressing all these aspects, apparent redundancy might still harbor hidden dependencies. For instance, if both flight computers use the same sensor inputs without cross-validation, a sensor failure could affect both systems despite their apparent independence.

2.2.5 Independence of Failures

Failure independence evaluates the likelihood of one failure affecting another within a system. This analysis is crucial for:

- Identifying single points of failure.
- Demonstrating that failures are unlikely to cascade across subsystems.

Ultimately, failure independence depends on rendering explicit all interfaces within the system. Cascading failures in apparently unrelated subsystems often come about through interfaces that were not obvious and/or not considered.

Example 2.6: *A water treatment plant has two supposedly independent chemical dosing systems:*

System A*: Controls chlorine levels for disinfection.* ***System B****: Controls pH levels through acid/base addition*

While these systems appear functionally independent, they share hidden dependencies:

(1) They both receive water flow rate data from a single flow meter. (2) They share a common PLC for control logic. (3) Their chemical supply lines run through the same containment area.

A failure analysis reveals that a single water leak in the containment area could damage both chemical supply systems, creating a common-mode failure. Similarly, an erroneous flow meter reading would cause both systems to dispense incorrect chemical amounts. True failure independence would require eliminating these shared dependencies or adding mitigations like separate flow meters and physical separation of supply lines.

2.3 Visualizing Independence (Orthogonal Views)

We use “views” as *axes* (what is separated and how we see it), while “types” describe *organizational forms* that realize those separations in practice.

So what does “good independence” look like when compared to “bad independence?” Visualizing independence allows us to examine and communicate the level of independence quickly and efficiently. Before generating a graph of the independence, we need to be clear on what we are trying to determine. Figure 2.3 maps these views onto orthogonal axes to make scope boundaries explicit.

The concept of independence can be visualized using orthogonal views in a three-dimensional (3D) knowledge space:

1. Domain Knowledge: Expertise in the application area of the system.
2. Design Knowledge: Understanding of the specific design under evaluation.
3. Process Knowledge: Awareness of the overall lifecycle and interdependencies.

Each individual involved in the process can be represented as a vector in this coordinate system, highlighting overlaps and gaps in independence. For example:

- A designer may score high in design knowledge but low in process knowledge.
- An auditor may have low design knowledge but high process knowledge.

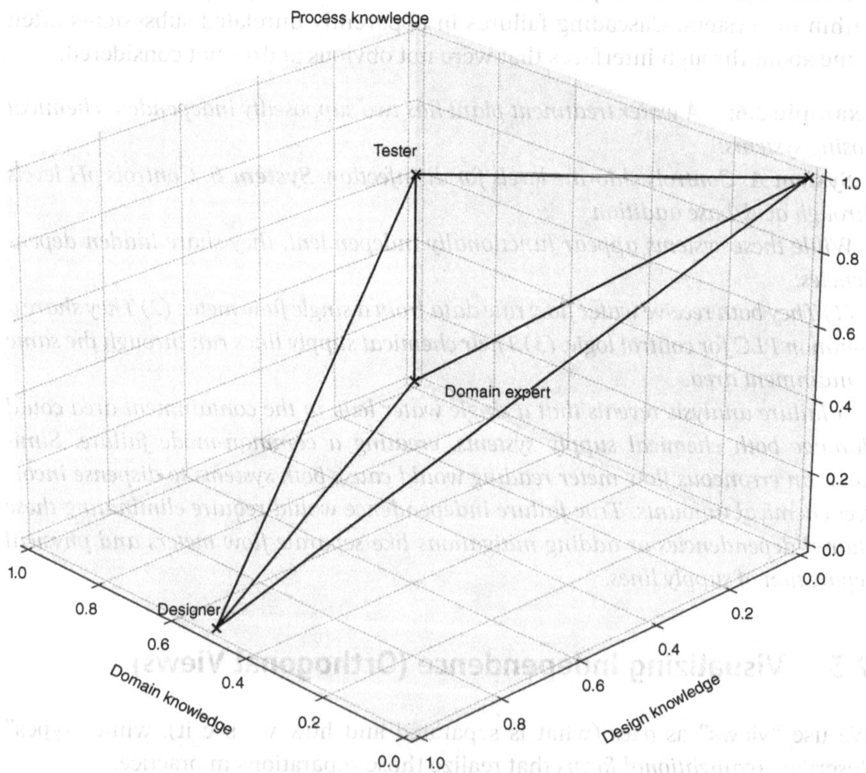

Figure 2.3 Orthogonal knowledge space for independence visualization.

If, in a project, we map the expertise of our resources onto a 3D space of their respective areas of expertise as a point, when we observe the points, they should be well distributed with minimal overlap on any of the axes—in other words, they are orthogonal. These visualizations help allocate resources effectively and maximize the impact of domain experts.

In testing, orthogonal views reveal how different testing approaches provide complementary coverage:

- Functional testing requires high domain knowledge to verify that system behavior matches domain requirements
- Structural testing requires high design knowledge to ensure code coverage and implementation correctness
- Process testing (such as reviewing development artifacts and procedures) requires high process knowledge to verify compliance with methodology

A testing strategy that combines these approaches with appropriate personnel will achieve greater independence and more comprehensive verification than relying on a single approach or homogeneous team composition.

2.4 Applying Independence

Independence is a practical tool for:

- Enhancing Robustness: Independent views ensure that errors are identified and addressed from multiple perspectives. The probability that two independent perspectives converge on the same, incorrect conclusion is significantly less than the probability that a single view converges on an incorrect conclusion.
- Optimizing Resources: Deploying experienced engineers on independent tasks maximizes their impact.
- Managing Risks: Understanding the degree of independence in processes informs project scheduling and risk management. Specifically, when independent processes come together (e.g., design and testing), it is a point where risk is eliminated. Conversely, until that time, you are carrying risk.

A simplified process perspective is shown in Figure 2.2, highlighting decision and evidence handoffs.

Example 2.7: *A medical imaging system development project applies independence across multiple dimensions:*

Process Independence: *Requirements analysis, design, implementation, and testing are performed by separate teams with different reporting structures. Each team uses the requirements specification as their baseline.*

Analysis Independence: *Both formal verification methods (mathematical modeling) and empirical testing (phantom studies and clinical evaluation) are used to validate image quality and radiation dose.*

System Independence: *Diagnostic and safety functions use separate processing chains and different technologies.*

When a subtle image artifact is missed by the development team but caught during independent clinical evaluation, this validates the independence approach. The artifact might have gone undetected if the same team performed both development and evaluation, as they would likely repeat the same blind spots.

By carefully defining, demonstrating, and applying independence, safety-critical systems can achieve greater reliability and resilience.

2.5 Too Much Independence

Example 2.8: *A critical design change to a braking algorithm is reviewed by an "independent" V&V team that reports to a separate division and cannot access near-real-time anomaly data. The team requests sanitized reports via a ticketing system with a 10-day SLA. During this delay, the design team proceeds, assuming "no objections." When the independent review finally flags an assumption mismatch (sensor drift model), the change is already integrated and tested in a larger build, making rollback risky and costly. Independence here reduced conflict of interest but also removed timeliness and context, degrading safety decisions.*

Although it is not a popular opinion, it is possible to have too much independence in an analysis. Too much independence shows up in analysis as "growing analyses"—which are analyses that are much larger than the item which is being analyzed. In practice, teams sometimes over-separate, treating their analyses as definitive and beyond challenge.

When a safety team begins to show signs that it believes that it is autonomous and completely divorced from the company to which it belongs is when a project is in real danger. This attitude of autonomy can manifest itself in numerous ways; the following are some symptoms that have been linked to excessive independence in the past:

- Appeal to Doctrine: Analyses are less technical than perhaps they should be and mix safety jargon with abstract safety principles without saying anything specific about the system that is being analyzed. E.g., "the system doesn't make use of sufficient elements that would lead to the maintenance of the required tolerable hazard rate (THR)." Blanket statements such as this example that don't add any quantitative data to the discussion are opinions that need to either be justified, quantified or bounded in some way. Much more appropriate would have been "The following fault chains were found that require only a single failure to result in a hazard..."—such a comment is focused, does not expand the arena of discussion and provides a clear area on which we should focus our efforts to improve the system.

- Acting Above the Rules: The safety team must adhere to the same rules as the rest of the engineering team. This means that if rules already exist for an element, then the safety team must either follow them, update the rules to be better or justify not following the rules—exactly like every other member of any other team. For example, when recommending additional requirements for a safety system, those requirements should be structured as valid requirements using the same format as the rest of the engineering team—an inability to structure a clear concise requirement is an indication that the thoughts surrounding it are unclear.
- Impractical or Overly Specific Requests: Safety teams could inadvertently make requests of the project that simply aren't feasible. Unfortunately, this could be a symptom of the safety team viewing itself as an autonomous entity that doesn't have to follow the same rules as the rest of the project team. Another option is that the safety team is less experienced in the specific area and didn't seek clarification. For example, is it practical for a safety team to request that an analysis be performed for all possible miscablings of a system? There are no doubt, situations where this is possible and/or required; however, two things need to be considered relative to this request:
  1. What are we actually trying to achieve?
  2. Is this really the only mechanism that is available to achieve the above?

  A safety team might be lacking experience in some of the practical aspects of building and deploying systems, so they might not be aware that the cabling is all built in a factory where it is tested with connectors that have unique mechanical keying already installed, then installed in the system without modification and tested after installation. Given this process, what would an analysis of all possible miscablings achieve? Are we trying to see what would happen when wire insulation fails or moisture enters into the system? The safety team's premise of answering the questions that were never asked is the problem. The requirement should have been something like: "The cabling of the system shall be performed in a manner so as to reduce or eliminate the possibility of the wiring being incorrect after installation and maintenance." This is still a poor requirement, in that it makes reference to the absolute (eliminate), but this will at least allow for a meaningful conversation around the specific topic.

- No consideration of the context: Again, typically an indication of an inexperienced safety team, an appeal to a mechanism or failure mode that is unnecessary given the specific operating context. This isn't something that should be discouraged; rather it should be encouraged that, when considering such elements of the safety analysis that the reasons that they do not need to be considered are included in the discussion.
- Inability to have their ideas challenged: Safety at its heart is a process that sometimes requires that ideas be challenged. A team that cannot accept having their ideas challenged could have this attitude because they have been operating autonomously rather than independently. Note that the design team having the same attitude is just as bad if not worse.

Example 2.9: *A railway signaling project illustrates the problem of excessive independence. The safety assessment team, operating with minimal interaction with the development team, produced a safety analysis that identified 127 "critical safety issues" requiring resolution before system approval.*

Upon closer examination, the project manager found that:

- *43 issues related to theoretical failure modes already mitigated by industry-standard practices*
- *35 issues were based on outdated system documentation*
- *28 issues used different terminology than the project standard, creating apparent (but not actual) problems*
- *Only 21 issues represented genuine safety concerns requiring attention*

The excessive independence of the safety team—operating in isolation rather than in constructive tension with the development team—created unnecessary work, schedule delays, and damaged credibility. A better approach would have maintained independence of perspective while establishing regular touchpoints to ensure accurate understanding of the system under assessment.

2.6 Balancing Independence and Integration

The ideal approach to independence is not isolation but rather a balanced integration that preserves diverse perspectives while enabling effective collaboration. Achieving this balance requires:

- Clear definition of independence objectives for each project phase
- Regular communication channels that preserve independence while preventing isolation
- Documentation of independence practices and their rationale

- Periodic review of independence effectiveness
- Leadership that values constructive tension

Independence should serve as a means to an end—improved safety, reliability, and quality—rather than an end in itself. When independence practices begin to hinder rather than help these goals, they should be reconsidered and refined.

Key Takeaways

- The key types of independence—process, analysis, organizational, system, and failure—provide different but complementary benefits in a safety-critical development process.
- By visualizing independence through orthogonal knowledge spaces, teams can better allocate resources and ensure complementary perspectives throughout the development lifecycle.

Part II

System Design

3

Engineering Change Control

After reading this chapter, you should be able to:

- Run a change through request, impact, decision, implementation, checks, and close-out.
- Write a short, clear change request with options and supporting evidence.
- Do a quick impact check across safety, interfaces, operations, and compliance.
- Link post-change checks to the stated risks.

Engineering change control (ECC) is a process for managing modifications to system designs, specifications, documentation, and procedures. In safety-critical systems, ECC ensures changes are introduced safely and transparently, reducing risks and maintaining compliance. ECC keeps system integrity, reliability, and safety intact throughout a system's lifecycle.

3.1 What Is Engineering Change Control?

For lifecycle context, see Section 5.4.

Engineering change control is the formal procedure through which proposed changes to system components, documentation, or processes are evaluated, approved, implemented, and tracked. ECC covers technical aspects (e.g., hardware and software modifications), procedural updates, and documentation revisions, ensuring that every modification is assessed for its impact on overall system safety, performance, and reliability. See Figure 3.1 for a lifecycle view of how changes are proposed, analyzed, approved, and verified.

Figure 3.1 Engineering change control lifecycle.

Key components of ECC include the following:

- **Identification of Changes:** Clearly defining what needs to be changed, why the change is necessary, and the expected benefits.
- **Impact Assessment:** Analyzing potential effects of changes on system performance, safety, and compliance, including direct and indirect impacts across related systems and components.
- **Approval Workflow:** Structured reviews and approvals by designated authorities with appropriate expertise and authority levels based on the change's risk classification.
- **Implementation Planning:** Detailed planning of how changes will be integrated, including resource allocation, scheduling, testing requirements, and contingency plans.
- **Documentation and Traceability:** Comprehensive record-keeping for transparency and compliance, ensuring all changes are traceable to requirements and approval decisions.

Example 3.1: *Consider a medical infusion pump that needs a software update to fix a timing issue. Under ECC:*

Identification: *The change request specifies updating the timing algorithm in the control software from version 2.1 to 2.2 to prevent delayed dose delivery during battery switchover.*

Impact Assessment: *Analysis shows the change affects the timing module, battery management interface, and alarm system. Testing must verify no regression in these systems and proper integration with existing hardware.*

Approval: *The CCB, including the software engineering lead, safety officer, and regulatory compliance manager, reviews and approves the change after confirming the impact assessment is complete and mitigation plans are adequate.*

This structured approach ensures the change is properly evaluated before implementation, maintaining the device's safety and integrity.

3.2 How to Implement Effective ECC

An ECC process needs clear procedures, good documentation, and communication among stakeholders.

3.2.1 Step-by-step Process

1. **Change Request (CR):** Document the proposed change, including justification, description, and initial assessment. The CR should state the business or technical need driving the change and provide enough detail for preliminary evaluation. Figure 3.2 provides a compact checklist for classifying impact across safety, cost, and schedule.

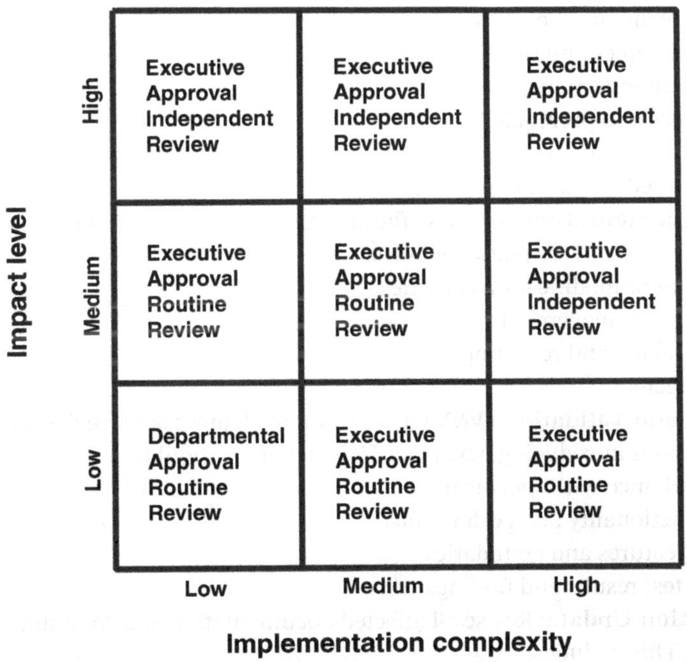

Figure 3.2 Change impact assessment matrix.

2. **Preliminary Review:** Conduct an initial screening to ensure completeness and clarity of the CR. This review also classifies the change based on its potential impact (e.g., minor, significant, major) to determine the appropriate approval path and level of scrutiny required.
3. **Detailed Impact Assessment:** Evaluate the change's effects on system safety, functionality, costs, schedules, and regulatory compliance. This assessment should include:
   - Technical impact analysis
   - Safety and hazard analysis
   - Resource requirements assessment
   - Schedule implications
   - Regulatory and compliance considerations
   - Risk assessment and mitigation planning
4. **Approval or Rejection:** A Change Control Board (CCB), composed of technical and safety experts, reviews and decides based on assessment outcomes. The CCB's composition may vary depending on the change's classification, with higher-impact changes requiring more senior representation and possibly external review.
5. **Implementation Planning:** Develop a detailed plan covering resources, schedule, and verification methods. The plan should include:
   - Specific implementation steps
   - Required resources and tools
   - Timeline with milestones
   - Dependencies and prerequisites
   - Verification and validation approach
   - Rollback procedures in case of issues
6. **Change Implementation:** Execute the approved changes following the implementation plan. This phase includes:
   - Development or modification activities
   - Configuration management
   - Progress tracking and reporting
   - Issue management
7. **Verification and Validation (V&V):** Confirm that changes achieve desired outcomes without introducing new risks. V&V activities should:
   - Verify compliance with specifications
   - Validate functionality and performance
   - Test safety features and boundaries
   - Document test results and findings
8. **Documentation Update:** Revise all affected documentation and communicate changes. This includes:
   - Technical documentation
   - User manuals and guides

   - Training materials
   - Regulatory submissions
   - Configuration records
9. **Post-implementation Review:** Evaluate effectiveness and identify lessons learned to inform future ECC activities. This review should assess:
   - Achievement of objectives
   - Adherence to plan and budget
   - Effectiveness of the ECC process itself
   - Areas for improvement

3.2.2 Roles and Responsibilities

Clearly defining roles ensures accountability and effective execution of the ECC process:

- **Change Originator:** Identifies and documents the proposed change. Responsible for providing clear justification and sufficient detail to enable proper evaluation.
- **Impact Analyst:** Conducts detailed impact assessment, ensuring all risks and benefits are clearly understood. This role requires both technical expertise in the affected systems and understanding of the broader operational context.
- **Change Control Board (CCB):** Reviews, approves, or rejects proposed changes, ensuring decisions are balanced and risk-informed. The CCB typically includes representatives from:
  - Engineering/Technical departments
  - Safety assurance
  - Quality management
  - Regulatory compliance
  - Project management
  - Operations (when applicable)
- **Implementation Team:** Executes approved changes, adhering strictly to approved plans and guidelines. This team must maintain detailed records of all implementation activities and report any deviations or issues promptly.
- **Verification and Validation Team:** Validates that implemented changes maintain or enhance safety and compliance, preventing unintended consequences. This team should be independent from the implementation team to ensure objective assessment.

Example 3.2: *For a railway signaling system software update:*

Change Originator: *A software engineer identifies a more efficient algorithm for train separation calculations that could reduce processing time by 15%.*

Impact Analyst: *A systems engineer evaluates how the algorithm change affects other system components, including interfaces with hardware, response times during fault conditions, and compatibility with existing safety mechanisms.*

CCB: *A board comprising the head of signaling engineering, safety certification manager, software quality lead, and operations representative reviews the impact assessment and approves the change with conditions for additional testing of boundary conditions.*

Implementation Team: *Software developers implement the new algorithm according to the approved specifications, with code reviews and unit testing.*

V&V Team: *Quality assurance engineers from a separate department conduct independent testing, including integration tests, performance under load, and verification against safety requirements.*

This separation of roles ensures multiple perspectives are considered, and no single person can implement changes without proper oversight.

3.3 Why ECC Matters in Safety-critical Systems

ECC is needed to maintain safety standards in complex systems.

3.3.1 Safety Assurance

Safety-critical systems cannot tolerate undocumented or uncontrolled changes, as minor modifications can lead to catastrophic failures. ECC ensures thorough review, documentation, and validation of changes, protecting system integrity and public safety.

Example 3.3: *Consider a change to a chemical processing control system that modifies a valve timing parameter. Without proper ECC:*

An engineer might adjust the parameters based on production efficiency calculations without considering that the faster valve closure could create pressure spikes in adjacent pipe sections.

With ECC in place:

- *The impact assessment identifies potential pressure spike risks*
- *Safety analysis determines maximum safe closure rates*
- *Testing verifies pressure remains within safe limits under various conditions*
- *Documentation ensures future modifications consider this constraint*

The ECC process prevents a seemingly minor efficiency improvement from compromising system safety by ensuring comprehensive evaluation of all potential consequences.

3.3.2 Compliance

Regulatory bodies mandate ECC processes to guarantee continuous adherence to safety standards. ECC serves as evidence of diligent management practices needed for audits and certifications.

3.3.3 Risk Management

ECC systematically evaluates risks associated with changes, allowing early identification and mitigation, reducing the likelihood of unintended consequences and failures.

3.4 ECC in Project Management

ECC controls and manages modifications within projects. The trade-off between change cost and risk exposure is sketched in Figure 3.3.

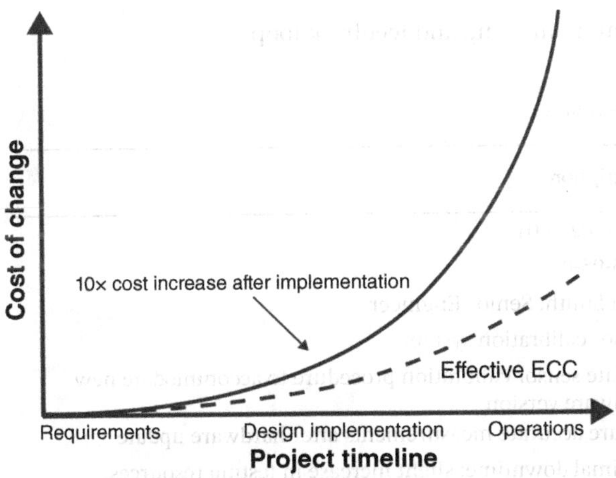

Figure 3.3 Change impact on project costs over time.

3.4.1 Project Stability

Uncontrolled changes disrupt project timelines and budgets. ECC brings predictability, allowing accurate planning and resource allocation.

3.4.2 Cost Control

ECC identifies potential costs early, allowing informed decision-making and reducing costly rework.

3.4.3 Communication and Transparency

ECC promotes transparency by documenting decisions and changes clearly, building stakeholder trust and reducing miscommunications.

3.5 Templates for ECC Documentation

See Table 3.1 for a concise change request template you can adapt.

Structured templates help with consistency, thoroughness, and documentation review.

3.6 Managing Emergency Changes

See Section 11.1 for incident monitoring and feedback loops.

Table 3.1 Change request template.

Field	Description
Change ID	ECC-2025-001
Date	2025-05-01
Originator	John Smith, Senior Engineer
System/component	Sensor calibration system
Description	Update sensor calibration procedure to accommodate new hardware version.
Justification	Ensure accurate measurements after hardware update.
Preliminary impact	Minimal downtime; slight increase in testing resources.
Change classification	Minor/significant/major (select one)
Attachments	List supporting documents, diagrams, etc.

Use Table 3.2 during triage to make safety and scope assumptions explicit.

Table 3.2 Impact assessment template.

Area	Impact	Mitigation required	Responsible party
Safety	High	Additional validation tests for sensor accuracy under extreme conditions	V&V team
Cost	Moderate	Budget adjustment of approx. US$15,000 for additional testing equipment	Project manager
Schedule	Low	Integrate into the planned maintenance window on June 15–16	Implementation team
Compliance	Medium	Update calibration documentation and recertify to ISO 13485 standards	Documentation lead
Training	Medium	Update training materials and conduct refreshers for maintenance staff	Training coordinator
Performance	Low	Monitor system response time during the first week after implementation	System engineer

Example 3.4: ***Emergency change with safety guardrails.*** *A field-critical patch disables a flaky sensor path and locks the system into a degraded but safe mode. The team executes a 15-minute impact huddle (single-owner decision, safety lead present), applies the patch behind a feature flag, logs the deviation, and schedules a 24-hour retrospective with full evidence collection. The emergency path is short and auditable, and it returns the system to the normal ECC process as quickly as possible.*

While structured processes matter, safety-critical environments sometimes require urgent changes to address critical issues. An ECC system must include provisions for emergency changes while maintaining necessary controls.

Key elements of emergency change management include:

- **Streamlined Approval Process:** Designated authorities who can approve emergency changes outside the standard CCB meeting cycle.
- **Minimum Documentation Requirements:** Simplified but sufficient documentation to capture essential information before implementation.
- **Post-implementation Review:** Thorough review after implementation to ensure proper documentation and control are reestablished.
- **Risk Assessment:** Rapid but structured assessment of potential impacts, with focus on primary safety concerns.

Example 3.5: *At a nuclear power plant, operators detect unexpected vibrations in a cooling pump. The situation doesn't warrant an immediate shutdown but requires prompt attention to prevent potential damage. The emergency ECC process allows:*

1. The shift supervisor and on-call engineer to approve a temporary reduction in pump speed based on an abbreviated impact assessment focusing on core cooling capacity.

2. Implementation of the change with minimal documentation but mandatory verification of cooling system performance.

3. Within 24 hours, the full CCB conducts a retrospective review, properly documents the change, and determines whether additional modifications are needed for long-term operation.

4. The emergency change is tracked specially to ensure proper follow-up and integration into the system's documentation.

This process balances the need for prompt action with the requirements for proper system control.

3.7 Integrating ECC with Configuration Management

Engineering change control works hand-in-hand with configuration management (CM) to ensure system integrity. While ECC focuses on the process of evaluating and implementing changes, CM ensures that the system's components and documentation accurately reflect its intended design at any point in time.

Key integration points include:

- **Baselines:** Establishing and maintaining defined configuration baselines that changes are applied against.
- **Version Control:** Ensuring all documentation and software/hardware specifications are under version control and updated as part of the change process.
- **Configuration Identification:** Uniquely identifying system components and their versions affected by changes.
- **Configuration Status Accounting:** Tracking the implementation status of approved changes across the system.

Effective integration ensures that:

- The current system configuration is always known
- Changes are applied to the correct baseline
- Historical configurations can be recreated if needed
- Compliance with requirements can be demonstrated

Key Takeaways

- ECC is necessary for managing safety-critical systems, reducing risks associated with change, and ensuring regulatory compliance.
- Implementing structured ECC processes improves the reliability and safety of systems and project management outcomes.
- Key points from this chapter:
  - ECC provides a structured framework for safely managing modifications to safety-critical systems
  - ECC requires clear processes, defined roles, and good documentation
  - The impact assessment phase identifies potential risks before changes are implemented
  - ECC supports both safety assurance and project management
  - Templates and standardized documentation help with consistency and compliance
  - Special provisions for emergency changes allow flexibility while maintaining control
  - Integration with configuration management ensures system integrity throughout the change process

In the next chapter, we will explore how ECC interfaces with broader system safety management practices and regulatory frameworks.

4

Building a Safety-critical System

After reading this chapter, you should be able to:

- Draft a simple safety plan: scope, lifecycle, roles, and independence.
- Name the reviews you will hold and what must exist to pass each one.
- Pick a few simple metrics that tell you if assurance is working.
- Keep hazards, analyses, and tests linked in one place.

The construction of a safety-critical system involves multiple interrelated processes and design elements. Designers must consider all aspects from initial requirements to final verification, ensuring the system is safe, reliable, and functional. Iterative refinement is needed for demonstrable correctness and completeness.

Independent formal reviews are central to demonstrating system correctness, requiring detailed information for accurate assessments.

4.1 System Context

Understanding the operational environment matters. The **system context** delineates system boundaries and external interactions—including human, mechanical, and digital interfaces. Documenting this context grounds design decisions in realistic conditions and constraints. For instance, railway signaling must account explicitly for weather, electromagnetic conditions, operational scenarios, and maintenance environments.

Example 4.1: *Consider an automated insulin delivery system for diabetic patients. The system context must define:*

Boundaries: *Where the system begins and ends (e.g., includes the insulin pump, glucose sensor, and control algorithm, but excludes the insulin manufacturing process)*

Interfaces:

- *Human: Patient interaction, healthcare provider configuration*
- *Mechanical: Insulin reservoir connection, infusion set attachment*
- *Digital: Smartphone app communication, hospital system integration*
- *Environmental: Operating temperature range (10–40 °C), water resistance (IPX8)*

Operational Scenarios: *Normal daily use, exercise, sleeping, illness, travel across time zones*

External Conditions: *Electromagnetic interference from airport scanners, pressure changes during air travel, extreme humidity*

This explicit context definition allows designers to anticipate safety challenges like sensor accuracy during exercise (when blood flow patterns change) or pump reliability in humid environments, ensuring appropriate mitigations are built into the design.

Clear context definitions help evaluate component independence within systems, which matters for safety validation.

4.2 General Design Considerations

System design follows criteria such as **correctness**, **independence**, and **completeness**. Every step must explicitly show its inputs, outputs, and validation methods for correctness, independence, and completeness.

Implicit assumptions must be made explicit and systematically tracked for verification.

4.2.1 Step Independence

Step independence ensures each design phase is executed as a discrete, self-contained process. Considerations include resource availability, timing constraints, clearly defined task scopes, and value contributions impacting overall safety.

Example 4.2: *In the development of an aircraft navigation system, step independence requires that the requirements analysis phase must be completed and formally* *reviewed before detailed design begins. This prevents the common problem of design decisions driving requirements rather than vice versa.*

To ensure true independence:

- *The requirements team delivers a complete, approved requirements specification with full traceability to regulations*
- *This document becomes a controlled input to the design team*
- *Any identified issues requiring requirements changes must go through formal change control rather than informal adjustments*
- *Design decisions must explicitly reference requirement IDs, making it clear when something lacks requirements coverage*

This independence ensures that safety requirements drive the design rather than being retrofitted to match an existing design concept, which could compromise safety integrity.

4.2.2 Completeness Criterion

Completeness verifies if a design step meets intended objectives. For instance, requirements analysis is complete once all requirements are documented and allocated, preventing redundancy and providing a solid foundation.

4.2.3 Step Context

Step context clarifies the relevance of a step within the larger system design, complementing completeness by outlining subsequent usage of step outputs.

4.2.4 Relationship to Other Design Steps

Each step must clearly define inputs and outputs, ensuring alignment and preventing overlap between phases for smooth transitions.

4.3 Generic Process Steps

For operational monitoring details, see Section 11.1.

4.3.1 Requirements Analysis

Designers must clearly identify, prioritize, and allocate functional and safety requirements early, resolving potential conflicts to ensure clarity and consistency. Figure 4.1 sketches the validation checkpoints and their relation to earlier design decisions.

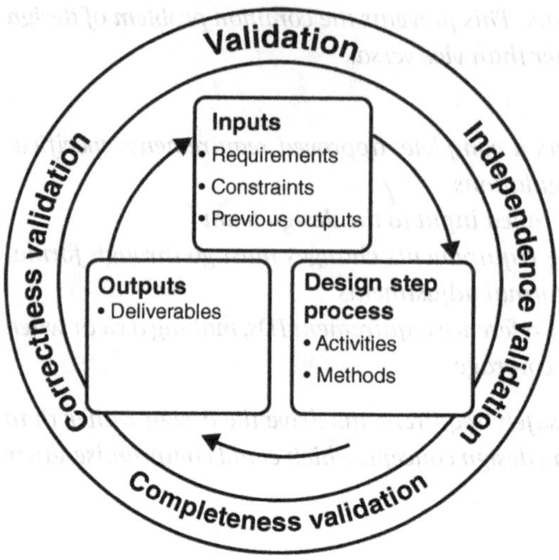

Figure 4.1 Design step validation framework.

Example 4.3: *For an emergency braking system in an autonomous vehicle, requirements analysis must balance potentially conflicting requirements:*

Functional Requirement: *"The system shall bring the vehicle to a complete stop within the shortest possible distance when an obstacle is detected."*

Safety Requirement: *"The system shall not decelerate at a rate that would cause occupants to experience more than the occupant tolerance defined in the applicable vehicle safety standard (e.g., ISO 26262 guidance)."*

Conflict: *The shortest possible stopping distance might require deceleration rates that exceed safety limits for occupants.*

Resolution: *"The emergency braking system shall bring the vehicle to a complete stop within the distance and deceleration limits specified by the applicable road-vehicle safety standard. Under imminent collision conditions, the system may temporarily exceed the nominal deceleration limit if this is required to avoid impact and still meets the standard's occupant tolerance." (Numbers are intentionally omitted; use domain standards and document the source.)*

This resolution establishes clear, measurable criteria that balance competing concerns and can be verified through testing.

4.3.2 Design for Analyzability

Modular designs, clear data and control flows, and simplified interactions help with hazard identification and system verification.

4.3.3 Managing Complexity

Complexity management involves simplicity, standardized components, and minimizing subsystem interdependencies, making validation easier.

4.3.4 Security Integration

Integrate cybersecurity proactively to mitigate threats that could compromise safety, using secure communications and strict access controls.

4.3.5 Design for Testability

Systems designed for testing make unit and integration tests easier, allow simulated inputs, and enable repeatable verification through deterministic designs.

4.3.6 Self-checks

Self-diagnostic systems, periodic health checks, watchdog timers, and redundancy improve reliability.

4.3.7 Latent Failure Detection

Design proactive mechanisms for identifying hidden faults, using predictive maintenance and rigorous testing to uncover latent failures.

4.3.8 Component Failure Modes

FMEA identifies component failure modes, guiding selection and integration of predictable components and fail-safe designs.

4.4 The Engineering Design Process

For independence concepts, see Section 2.1.

1. **Requirements Analysis:** Validate actionable, testable safety requirements. See Figure 4.2 for a high-level view of how development and assurance activities interact over the lifecycle.
2. **Systems Engineering:** Define robust architecture, clearly managing subsystem interfaces.
3. **Hardware Engineering:** Select reliable components, design redundancy, address environmental challenges.

4. **Software Engineering:** Adhere to coding standards (e.g., MISRA C:2012 Guidelines for the Use of the C Language MISRA [2]), verify correctness formally, and employ defensive programming.
5. **Test Engineering:** Comprehensive testing plans, automation, and failure scenario simulations.

Example 4.4: *In developing safety-critical software for a medical radiation therapy system, defensive programming techniques include:*

Input Validation: *Before calculating radiation dose, the software validates that:*

- *Patient parameters are within physiological ranges (e.g., weight between 10 and 300 kg)*
- *Requested radiation dose is within equipment capabilities*
- *Treatment area coordinates fall within calibrated machine boundaries*

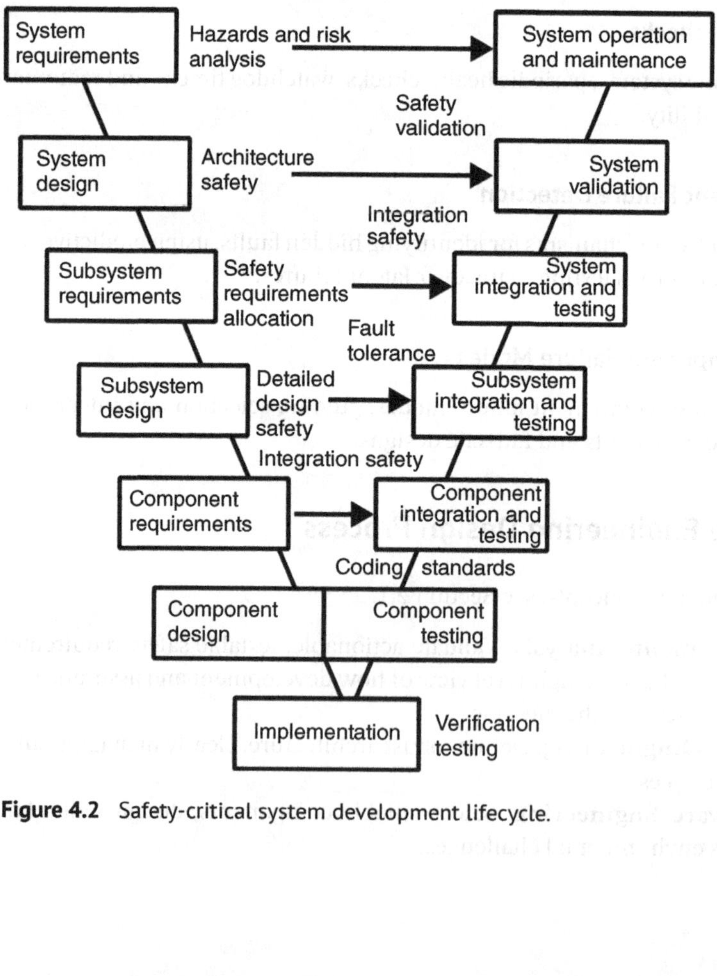

Figure 4.2 Safety-critical system development lifecycle.

Runtime Assertions: *During calculation, the software continuously checks:*

- *Intermediate results remain within expected ranges*
- *Computation time doesn't exceed safe limits*
- *Hardware sensors report consistent readings*

Error Handling: *If any check fails, the software:*

- *Immediately stops radiation delivery*
- *Records detailed diagnostic information*
- *Issues clear alerts to operators*
- *Requires explicit override to continue*

These techniques create multiple layers of protection against programming errors, unexpected inputs, or hardware failures that could lead to patient harm.

4.5 Engineering Techniques

4.5.1 Independent Monitoring

Use independent monitoring modules to continuously verify system behavior and maintain fail-safety.

4.5.2 Latent Failure Detection Techniques

Use real-time diagnostics and predictive techniques (including machine learning) for failure anticipation.

4.5.3 Functional Segmentation

Segregate critical and non-critical functions, isolating and containing failures.

Example 4.5: *In an industrial control system for a chemical processing plant, functional segmentation divides the system into isolated domains with different criticality levels:*

Critical Safety Functions (Highest Criticality):

- *Emergency shutdown systems*
- *Pressure relief monitoring*
- *Toxic gas detection*

These run on dedicated hardware with simplified, formally verified software and have priority access to all actuators.

Process Control Functions (Medium Criticality):

- *Temperature regulation*
- *Flow control*
- *Mixture proportioning*

These run on separate hardware from safety functions and cannot override safety systems.

Administrative Functions (Low Criticality):

- *Production reporting*
- *Efficiency optimization*
- *Maintenance scheduling*

These are isolated on separate networks with no direct control capabilities.

This segmentation ensures that a failure or compromise in lower-criticality functions (e.g., a reporting system crash) cannot affect safety-critical operations, while safety functions remain simple enough for comprehensive verification.

4.5.4 Compound Actuators

Implement actuator redundancy to improve reliability through diversity in failure characteristics.

4.5.5 Addressing Known Failure Modes

Integrate fault-tolerant components and predefined mitigation strategies for known failure scenarios.

4.6 Safety Architectures

For the verification stages referenced here, see Section 5.4.

Example 4.6: ***Design Trade-off.*** *A dual-channel braking controller considers two architectures: (1) symmetric channels with cross-check and majority voting; (2) an active primary with a simpler independent monitor that asserts a safe state on threshold violations. (1) offers graceful degradation but raises common-mode complexity; (2) simplifies verification and fault containment but requires careful threshold definition and diagnostics to avoid nuisance trips. The choice depends on hazard severity, fault detectability, and verification budget.*

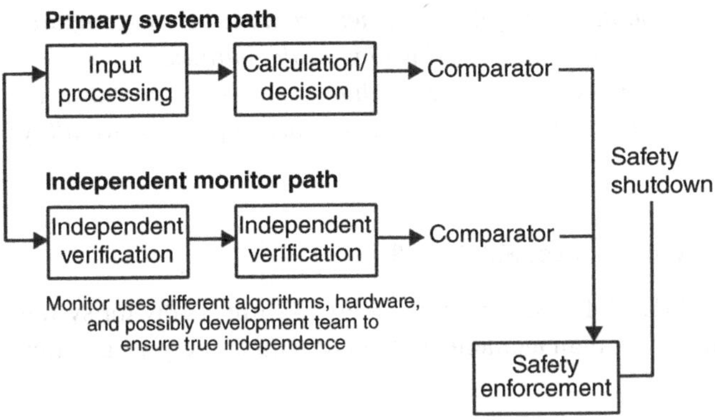

Figure 4.3 Independent monitoring architecture.

4.6.1 Inherently Fail-safe Designs

Systems default safely upon failure; e.g., railway signals default to "stop" conditions during faults. Figure 4.3 illustrates a common independent-monitoring pattern and where it fits in the architecture.

4.6.2 Composite Safety

Combine redundancy, diversity, and self-checking to address failures comprehensively.

Example 4.7: *A flight control system for a commercial aircraft employs composite safety through multiple strategies:*

Redundancy: *Three identical flight control computers operate in parallel, each capable of controlling the aircraft.*

Diversity: *Each computer uses different:*

- *Hardware (different processor architectures from different manufacturers);*
- *Software (independently developed by separate teams);*
- *Power sources (primary, secondary, and emergency power).*

Self-checking: *Each computer:*

- *Continuously monitors its internal operations;*
- *Cross-checks its outputs against the other computers;*
- *Uses built-in test equipment to verify sensor inputs.*

Voting: *Control surface commands require agreement from at least two computers, with the third computer's disagreement triggering diagnostic routines.*

This composite approach ensures that no single failure point—whether hardware, software, development error, or systematic design flaw—can compromise overall system safety.

4.6.3 Standards and Compliance: MIL-STD 882

Apply the principles of MIL-STD-882E: System Safety [3] universally for systematic hazard identification, risk management, and documented safety verification.

4.6.4 Achieving Safety Through Reliability

Reliability-driven safety relies on validated, dependable components supported by rigorous testing.

4.7 Safety Lifecycle Management

While the previous sections focused on design aspects, safety must be managed throughout the entire system lifecycle.

4.7.1 Verification and Validation Strategy

Develop a strategy for verifying that the system is built correctly and validating that it meets safety requirements. This strategy should:

- Define verification methods appropriate for each requirement
- Establish independence between development and verification teams
- Include coverage analysis to ensure completeness
- Provide traceability from requirements to verification evidence

4.7.2 Configuration Management

Implement rigorous configuration management practices to maintain control over:

- Design documentation
- Software and hardware versions
- Test procedures and results
- Safety analyses
- Safety case evidence

4.7.3 Change Management

Establish formal processes for evaluating and implementing changes, ensuring that:

- Safety impacts are thoroughly assessed
- Regression testing confirms continued safety
- Documentation remains current
- Approvals are obtained at appropriate levels

4.7.4 Operational Monitoring

Design for ongoing safety monitoring during operational use, including:

- Incident reporting systems
- Usage data collection
- Performance monitoring
- Regular safety assessments.

Key Takeaways

- Constructing safety-critical systems requires following design principles, structured processes, and sound safety architectures.
- Addressing these areas well results in systems that meet safety standards and operate reliably and securely.
- Key points from this chapter:
  - Safety must be designed in from the beginning, not added later
  - Clear system context definition provides the foundation for safety design
  - Independence between design steps ensures objectivity and thoroughness
  - Multiple complementary techniques provide defense-in-depth against failures
  - Systematic processes reduce human error and ensure comprehensive coverage
  - Safety must be maintained throughout the entire system lifecycle.

Subsequent chapters will explore methodologies and tools that build on these elements, with particular focus on verification techniques and safety case development.

Part III

Process and Practice

5

Verification and Validation

After reading this chapter, you should be able to:

- Say what is verification and what is validation, with an example of each.
- Pick test and analysis methods that fit the risk and what you can observe.
- Write a simple V&V plan and name the pass/fail rules.
- Build a small RTM linking hazards, requirements, and tests.

The verification and validation (V&V) process ensures safety, functionality, and reliability in safety-critical systems. It provides systematic assurance that the system is constructed correctly (verification) and fulfills its intended purpose within its operational context (validation).

5.1 The "Checking" Process

The V&V process systematically addresses two questions that matter for safety and quality assurance: See Figure 5.1 for an overview of where verification and validation activities sit in the lifecycle.

- **Verification:** Are we building the system correctly?
  - Verification involves ensuring compliance with requirements, specifications, and standards at each development stage. Methods include rigorous reviews, inspections, audits, static and dynamic testing, formal code analysis, and formal verification techniques such as theorem proving and model checking.
  - Key activities include establishing requirements traceability, conducting peer reviews, performing unit and integration tests, and carrying out comprehensive system-level testing. Emphasis on early detection reduces risks associated with late-stage defect resolution.

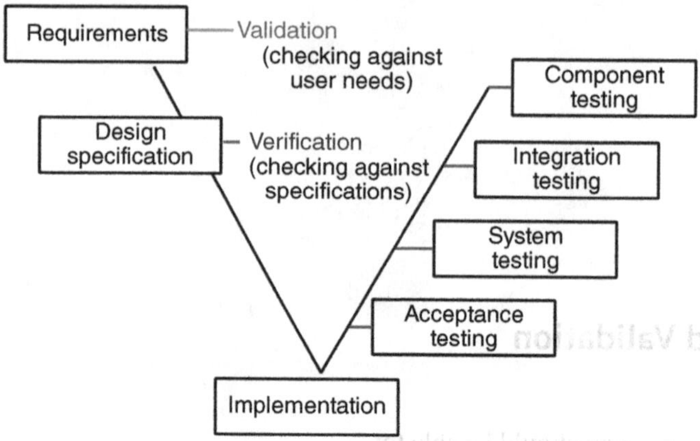

Figure 5.1 The V&V process in the system development lifecycle.

- **Validation:** Are we building the right system?
  - Validation ensures the system meets operational requirements and user expectations. It encompasses operational scenario testing, user acceptance tests, field trials, realistic operational simulations, and evaluations against documented operational requirements.
  - Effective validation requires engaging end-users and stakeholders early and continuously throughout development to ensure alignment with real-world usage conditions.

Example 5.1: *Consider the development of an automated external defibrillator (AED) for public use:*

Verification Activities *(building the system correctly):*

- *Checking that the electrical output matches the specified range (e.g.,* 150–360 J*)*
- *Testing that the device powers on within the required time frame (*<10 seconds*)*
- *Confirming that the software correctly implements the heart rhythm analysis algorithm according to its specification*
- *Verifying that the battery life meets the specified five-year standby requirement*

Validation Activities *(building the right system):*

- *Testing whether untrained users can successfully operate the device in simulated emergency conditions*
- *Evaluating if the voice instructions are clear and effective in high-stress situations*
- *Confirming the device can function properly in various environmental conditions (temperature, humidity, background noise)*
- *Assessing if the device achieves the ultimate goal of helping users successfully resuscitate patients in cardiac arrest*

This example illustrates that verification focuses on confirming the device meets its technical specifications, while validation ensures the device fulfills its intended purpose in its actual usage context.

5.2 Key Principles of the "Checking" Process

V&V activities follow core principles to be effective:

- **Independence:**
  - Verification and validation must be conducted by personnel independent from the system's designers or implementers, reducing bias and promoting objective assessments.
  - Independent teams, internal or external, ensure unbiased evaluation and add credibility to outcomes.
- **Iterative Approach:**
  - Implementing V&V iteratively throughout the development lifecycle enables early detection and timely resolution of issues, preventing cumulative errors and late-stage defect discovery.
  - Often overlooked, early iterative planning and structured impact analysis reduce the risk of undiscovered safety-critical errors.
  - Incorporating regular impact analyses during early planning phases ensures systematic management of system modifications, updates, and incremental improvements.
- **Documented Evidence:**
  - Thorough documentation of V&V processes provides evidence for compliance with safety and regulatory standards and supports traceability and transparency.
  - Key documentation includes V&V plans, detailed test procedures, traceability matrices, comprehensive test results, inspection and audit reports, risk assessments, and documented rationale for deviations or exceptions.
- **Tailored Rigor:**
  - The level of V&V rigor should be proportional to the criticality of the component or system being evaluated.
  - Safety-critical functions require more extensive testing, formal methods, and analysis than non-critical functions.
  - Resource allocation should prioritize areas of highest risk or greatest impact on system safety.

5.3 Impact of Neglecting Early Iterative Planning

Many projects underestimate proactive iterative planning, only introducing iterative processes reactively. Such reactive planning leads to fragmented, incomplete impact analyses, resulting in superficial evaluations of system changes. Late-stage impact analyses can miss critical system dependencies, insufficiently evaluate regression testing needs, and overlook interactions that compromise system integrity.

Example 5.2: *Consider a train control system where a seemingly minor software update to improve energy efficiency is implemented without proper iterative planning:*

Without Early Iterative Planning:

- *Week 1: Software team modifies the acceleration algorithm to optimize energy use*
- *Week 6: Testing identifies that the modified algorithm sometimes delays emergency braking by* 0.5 seconds
- *Week 8: Engineers discover this delay violates safety requirements for stopping distance*
- *Week 12: Major redesign required, delaying project by months and requiring extensive recertification*

With Early Iterative Planning:

- *Before implementation: Impact analysis identifies all functions affected by acceleration changes*
- *Planning phase: Testing strategy includes specific verification of braking response time*
- *Initial design phase: Small prototype demonstrates potential braking delay, leading to algorithm adjustment*
- *Implementation phase: Modified design proceeds with safety requirements already satisfied*
- *Testing phase: Verification confirms both energy efficiency and safety requirements are met*

This example demonstrates how early iterative planning identifies potential issues before they become embedded in the system, saving time, reducing costs, and maintaining safety integrity.

Effective impact analysis requires early identification and systematic management of system changes, ensuring comprehensive safety considerations, reducing risks, and improving system reliability and compliance.

5.4 The V&V Lifecycle

For independence and governance context, see Section 2.1.

V&V is not a single phase but spans the entire system development lifecycle, with specific activities at each stage: Figure 5.2 highlights how independence evolves across the lifecycle and where review gates typically sit.

5.4.1 Requirements Phase V&V

- **Verification Activities**:
  - Requirements reviews to check for clarity, completeness, consistency, and testability
  - Formal methods to mathematically verify requirements properties
  - Requirements traceability analysis to ensure regulatory compliance

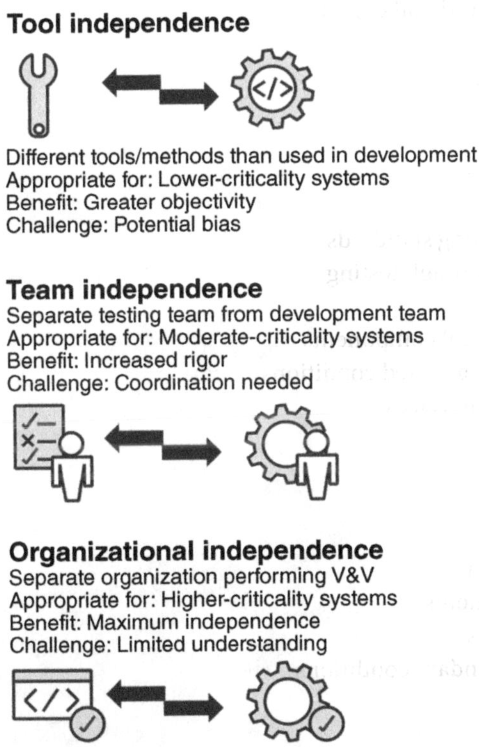

Figure 5.2 Independence in verification and validation.

- **Validation Activities**:
  - Stakeholder reviews of requirements
  - Prototyping and simulations to validate concept feasibility
  - Operational scenario analysis to verify requirements address real-world needs

5.4.2 Design Phase V&V

- **Verification Activities**:
  - Design reviews against requirements
  - Analysis of design models for compliance with standards
  - Formal verification of critical algorithms
  - Safety analyses such as FMEA and FTA
- **Validation Activities**:
  - Prototype evaluations with end-users
  - Simulation of system behavior under various operational scenarios
  - Design walkthroughs with operational experts

5.4.3 Implementation Phase V&V

- **Verification Activities**:
  - Code reviews and static analysis
  - Unit testing of components
  - Compliance checks against coding standards
  - Coverage analysis to ensure thorough testing
- **Validation Activities**:
  - Early integration testing of critical components
  - Performance evaluation under expected conditions
  - Usability assessments of user interfaces

5.4.4 Testing Phase V&V

- **Verification Activities**:
  - Integration testing of subsystems
  - System testing against requirements
  - Regression testing after changes
  - Performance testing under boundary conditions
- **Validation Activities**:
  - User acceptance testing
  - Operational testing in representative environments
  - Scenario-based testing of real-world conditions

5.4.5 Deployment and Maintenance V&V

- **Verification Activities**:
  - Installation testing
  - Configuration audits
  - Update verification
  - Continued compliance monitoring
- **Validation Activities**:
  - Post-deployment monitoring
  - User feedback collection
  - Operational performance analysis
  - Long-term safety assessment

5.5 Reducing Risk Through Effective V&V

Well-executed V&V processes reduce project risks by ensuring the defect-escape vs. cost curve in Figure 5.3 motivates early detection and tight feedback loops.

- Early defect detection and resolution.
- Compliance with regulatory and safety standards.
- Alignment of system functionality with operational expectations, reducing costly redesigns and schedule delays.
- Transparent documentation and thorough testing, which builds stakeholder confidence, helps with regulatory approval, and decreases lifecycle costs.

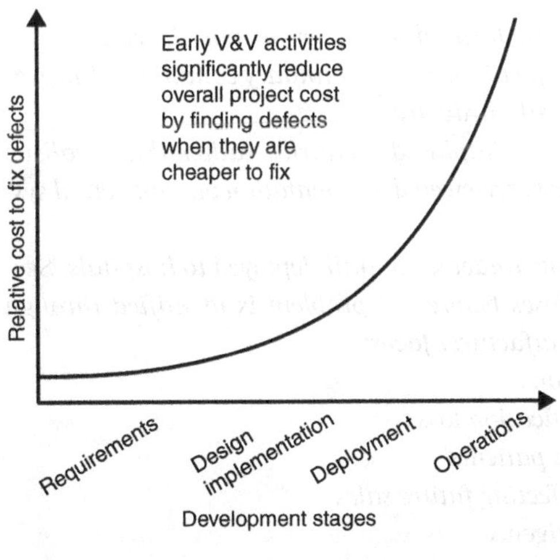

Figure 5.3 Cost of defect correction across development stages.

5.6 Implications of Improper V&V Execution

Example 5.3: ***Missed Boundary Condition.*** *A speed computation module passes unit tests but fails at an exact threshold (e.g., 100.0 km/h) due to a rounding assumption mismatch between module and integration code. The defect escapes to system test, delaying release. Mitigation: require explicit boundary-value tests in the component test plan, add property-based checks for monotonicity, and align numeric assumptions in the interface control document (ICD). A single assertion at the interface prevents recurrence.*

Improper or neglected V&V carries substantial negative implications, including:

- Increased likelihood of undetected defects, potentially causing safety incidents, system failures, or catastrophic outcomes.
- Regulatory non-compliance, risking financial penalties, legal liabilities, and reputational damage.
- Significant project schedule and cost overruns due to late-stage defect discovery and required rework.
- Erosion of stakeholder trust, diminished market credibility, and broader negative organizational impacts.

Example 5.4: *Consider a radiation therapy system where improper V&V execution has serious consequences:*

Software Defect: *A unit conversion error in the dose calculation algorithm occasionally results in incorrect radiation doses.*

Improper V&V Impact:

- ***Verification Failure:*** *Code reviews focused only on style, not algorithmic correctness. Unit tests only checked typical cases, not boundary conditions. Integration testing used simulated, not real hardware.*
- ***Validation Failure:*** *Testing used simplified scenarios rather than realistic treatment plans. Clinical specialists reviewed documentation but not actual system operation.*
- ***Consequences:*** *The error remains undetected until deployed to hospitals. Several patients receive incorrect doses before the problem is identified through adverse event reporting. The manufacturer faces:*
  - *Mandatory recall costing millions*
  - *Regulatory sanctions and certification loss*
  - *Multiple lawsuits from affected patients*
  - *Major reputational damage affecting future sales*
  - *Criminal investigation for negligence*

This example illustrates how inadequate V&V can allow critical defects to reach deployment, resulting in severe safety, financial, legal, and reputational consequences.

Real-world failures reinforce these points. Inadequate software checks and configuration management contributed to the Ariane 5 Flight 501 loss [4]; missing safeguards and weak test coverage led to overdoses in the Therac-25 incidents [5]; and insufficient independence and scenario coverage in certification contributed to the Boeing 737 MAX MCAS issues [6]. When citing incidents, point to the official investigation reports to anchor lessons in evidence.

Rigorous and timely V&V execution is needed to ensure safety, reliability, and overall project success.

5.7 V&V Documentation and Deliverables

Good documentation is needed for effective V&V processes. Key deliverables include the following.

5.7.1 V&V Plan

This master document outlines the overall approach to verification and validation, including:

- Scope and objectives
- Roles and responsibilities
- Resource requirements
- Schedule and milestones
- Methods, tools, and techniques
- Independence strategy
- Defect management process
- Reporting mechanisms

5.7.2 Requirements Traceability Matrix

This document maps requirements to verification methods and test cases, ensuring:

- Every requirement is verified
- Each test contributes to requirement verification
- Impact of changes can be assessed
- Gaps in testing coverage are identified

5.7.3 Test Documentation

Comprehensive test documentation includes:

- Test plans outlining testing strategy
- Test specifications detailing test cases
- Test procedures with step-by-step instructions
- Test reports summarizing results and issues
- Test data sets used for validation

5.7.4 Anomaly Reports

Documentation of deviations, defects, and anomalies found during V&V, including:

- Description of the issue
- Severity and priority
- Root cause analysis
- Resolution status
- Verification of corrections

5.7.5 V&V Final Report

A comprehensive summary of all V&V activities, including:

- Overall assessment of system quality
- Summary of verification results
- Summary of validation results
- List of open issues and their resolution
- Recommendations for system acceptance or deployment

Key Takeaways

- The verification and validation process is central to ensuring safety, reliability, and functionality in safety-critical systems.
- Subsequent chapters will explore specific verification and validation techniques in greater detail, providing practical guidance for their implementation across different domains.

6

Analytical Methods

After reading this chapter, you should be able to:

- Fill in a clean Failure Modes and Effects Analysis (FMEA) row and a small fault tree.
- Choose a method (FMEA, Fault Tree Analysis, HAZOP, STPA) and state why it fits.
- Turn analysis results into a requirement and a test.
- Review an analysis and note one gap to fix.

Safety-critical systems require rigorous analysis to systematically identify hazards, assess risks, and verify system reliability. Analytical methods give designers and analysts structured tools to evaluate systems early and make risk-based decisions. This chapter covers key analytical methods with clear explanations, detailed examples, and practical guidance for implementation. Examples are illustrative and walk through inputs, steps, and outputs so you can translate the pattern to your own domain.

6.1 Deterministic Analysis

For risk quantification framing, see Section 1.3.1.

Deterministic analysis evaluates system behaviors under predefined, known conditions, assuming certainty in input values and system states. See Figure 6.1 for a side-by-side contrast with probabilistic reasoning.

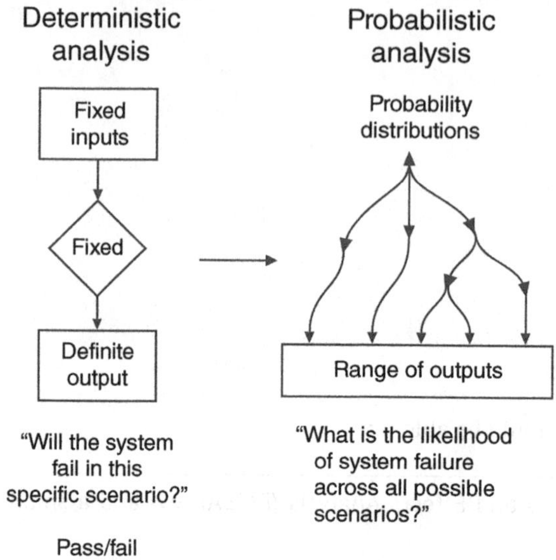

Figure 6.1 Deterministic vs. probabilistic analysis comparison.

6.1.1 Key Characteristics

Deterministic methods yield predictable, repeatable outcomes due to fixed inputs. They rely on clear logical reasoning and explicit system definitions, making them ideal for analyzing well-defined nominal and fault conditions.

Key attributes include:

- **Fixed Inputs and Outputs:** Given the same inputs, deterministic analyses always produce identical results
- **Clear Causality:** Direct cause-and-effect relationships are established and verified
- **Scenario-based:** Analysis focuses on specific, well-defined scenarios
- **Boundary Condition Focus:** Particular attention to system behavior at operational limits
- **Binary Pass/Fail Criteria:** Clear thresholds for acceptable system performance

6.1.2 Illustrative Example

Automated Fire Suppression System: Analysts examine how quickly sprinklers activate after smoke detection, establishing maximum allowable delay times to prevent the spread of fire. Analysts verify system behavior against clearly defined performance criteria, confirming compliance with safety requirements.

Example 6.1: *Consider a railway signaling system where a deterministic analysis focuses on the fail-safe behavior of signals:*

Setup: *The safety requirement states that if communication between a signal and the control center is lost for more than 3 seconds, the signal must automatically revert to "stop" (red).*

Deterministic Analysis Steps:

1. ***Define Test Scenarios:***
   - *Communication loss for 2.5 seconds, then restored*
   - *Communication loss for 3.5 seconds, then restored*
   - *Intermittent communication losses (multiple 1-second losses)*
   - *Gradual degradation of communication quality*
2. ***Create Test Cases with Exact Inputs:*** *For each scenario, precisely define initial signal state, timing of communication loss, and system configuration.*
3. ***Execute Tests:*** *Run each scenario through simulation or physical testing.*
4. ***Evaluate Against Binary Criteria:*** *For each test:*
   - *When communication loss <3 seconds: Signal must maintain its original state*
   - *When communication loss >3 seconds: Signal must change to "stop" within 0.5 seconds*
5. ***Document Results:*** *Record exact system response times and behavior for each scenario.*

This deterministic analysis verifies that the signaling system unfailingly meets its safety requirements under precisely defined conditions, providing certainty about the system's behavior in these specific scenarios.

6.1.3 Benefits and Limitations

6.1.3.1 Benefits

- **Clarity and Transparency:** Straightforward cause-and-effect relationships are easy to explain and justify to stakeholders
- **Precise Verification:** Enables exact verification of system behavior against specific requirements
- **Design Validation:** Provides concrete evidence that design specifications are correctly implemented
- **Reproducibility:** Tests can be repeated with identical results, supporting rigorous verification
- **Regulatory Acceptance:** Well-aligned with prescriptive regulatory frameworks that specify exact requirements

6.1.3.2 Limitations

- **Limited Scope:** Analyzes only predefined scenarios, potentially missing unexpected interactions
- **Uncertainty Exclusion:** Cannot account for real-world variability and uncertainties
- **Scenario Selection Bias:** Effectiveness depends on analysts' ability to identify relevant scenarios
- **Binary Thinking:** May oversimplify complex systems with nuanced performance characteristics
- **Resource Intensity:** Comprehensive coverage of all possible scenarios is impractical for complex systems

These strengths and weaknesses are not abstract. In practice, deterministic work shines when the team shares a crisp understanding of the system and its limits; it falters when the wrong scenario set is chosen. When you pick deterministic checks, write down why you picked them and what you are deliberately leaving out, so later readers understand the trade.

6.2 Probabilistic Analysis

For lifecycle placement of evidence, see Section 5.4.

Probabilistic analysis incorporates uncertainty into evaluations by using statistical and mathematical modeling, estimating the likelihood and impact of various system outcomes. Figure 6.1 highlights where frequency/likelihood modeling complements deterministic bounds.

6.2.1 Key Characteristics

This approach provides probability-based insights, highlighting risks of system failure or performance degradation by using historical data, statistical techniques, and stochastic processes.

Essential features include:

- **Uncertainty Modeling:** Explicitly represents variability in inputs, environments, and behaviors
- **Distribution-based:** Uses probability distributions rather than single values
- **Quantified Confidence:** Provides statistical confidence levels for analysis results
- **Risk Quantification:** Expresses safety risks in numerical probability terms
- **Long-term Prediction:** Enables forecasting of rare events over extended operational periods

6.2.2 Illustrative Example

Emergency generator reliability: Analysts calculate the probability that backup power systems fail during a blackout, considering historical failure rates, environmental conditions, and maintenance schedules. Probabilistic analysis produces failure likelihood estimates to inform maintenance strategies and redundancy requirements.

Example 6.2: *Consider a medical infusion pump delivering critical medication where probabilistic analysis is used to evaluate overall system reliability:*

Setup: *Instead of analyzing specific failure scenarios, probabilistic analysis examines the overall likelihood of pump failure under varying real-world conditions.*

Probabilistic Analysis Steps:

1. ***Identify Key Components and Their Failure Distributions:***
   - *Pump motor: Weibull distribution with parameters derived from testing*
   - *Battery: Exponential degradation based on charging cycles*
   - *Pressure sensors: Normal distribution for drift characteristics*
   - *Software: Defect density model for potential runtime errors*
2. ***Construct a System Reliability Model:*** *Using techniques such as reliability block diagrams (RBD) or Markov models to represent component interdependencies*
3. ***Run Monte Carlo simulation:*** *Perform 10,000+ simulations drawing random values from each component's failure distribution*
4. ***Analyze Reliability Metrics:***
   - *Calculate mean time between failures (MTBF): 8760 hours*
   - *Determine probability of failure during typical 48-hour patient treatment: 0.55*
   - *Identify probability of undetected failure: 0.03*
   - *Calculate 95% confidence intervals for all metrics*
5. ***Sensitivity Analysis:*** *Determine which components most significantly affect overall reliability*

The probabilistic analysis reveals that while deterministic testing showed the pump meets all specified requirements, there remains a quantifiable (though small) risk of failure during normal operation. This information guides decisions about maintenance intervals, redundancy needs, and monitoring requirements that wouldn't be apparent from deterministic analysis alone.

6.2.3 Benefits and Limitations

6.2.3.1 Benefits

- **Reality Alignment:** Better reflects real-world variability and uncertainty
- **Comprehensive Coverage:** Considers the full spectrum of possible scenarios and conditions

- **Risk Insights:** Provides a quantitative basis for risk-informed decision-making
- **Resource Prioritization:** Identifies highest-risk components and scenarios for focused attention
- **Rare Event Assessment:** Enables evaluation of low-probability, high-consequence events that deterministic testing cannot practically cover

6.2.3.2 Limitations

- **Data Dependency:** Requires substantial historical or test data for accurate probability distributions
- **Complexity:** Models and results may be difficult for stakeholders to understand
- **Interpretation Challenges:** Probabilistic results require careful explanation and contextual interpretation
- **Validation Difficulty:** Challenging to validate models against real-world behavior
- **Computational Demands:** Sophisticated probabilistic analyses may require significant computing resources

Probabilistic arguments feel less tangible than step-by-step proofs, so explain the story in plain words: what data you used, what confidence you actually have, and how the numbers change what you will do. This helps nonspecialists trust the result instead of dismissing it as "statistics magic."

6.3 Orthogonal Views

Orthogonal views integrate multiple independent analytical perspectives to ensure thorough system evaluation. Combining insights from different domains reduces the risk of overlooking interactions and dependencies that matter for system safety. Figure 6.2 provides a compact orientation to the analysis perspectives used throughout this chapter.

6.3.1 Analytical Dimensions

Orthogonal views integrate:

- **Domain Knowledge:** Understanding specific operational contexts and safety requirements.
- **Design Knowledge:** Detailed knowledge of system architecture, internal functionalities, and component interactions.
- **Process Knowledge:** Insights into system lifecycle processes, regulatory compliance, maintenance requirements, and procedural dependencies.

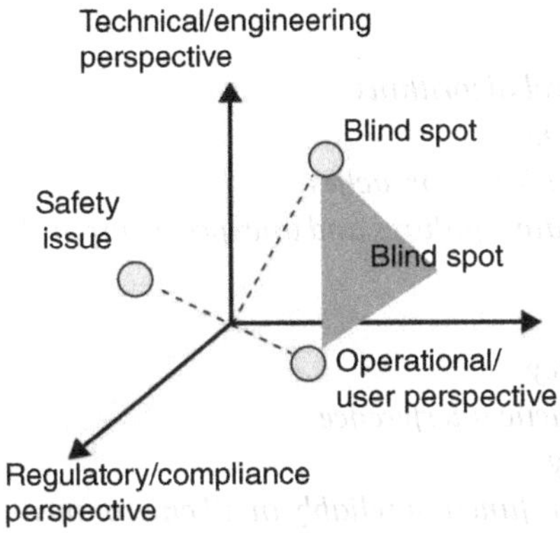

Figure 6.2 Orthogonal views analysis framework.

- **Human Factors Perspective:** Consideration of human-system interactions, usability, and human error potential.
- **Environmental Context:** Analysis of external environmental conditions and their effects on system operations.

6.3.2 Illustrative Example

Medical Device Safety Evaluation:

- **Domain Expert:** Evaluates patient safety implications, clinical usability, and compliance with medical safety standards.
- **System Designer:** Assesses internal system reliability, software-hardware interactions, and operational effectiveness.
- **Process Specialist:** Ensures compliance with regulatory frameworks, evaluates maintenance procedures, and lifecycle management processes.

Example 6.3: *Consider an aircraft flight control system analysis using orthogonal views:*

The System: *A fly-by-wire flight control system that translates pilot inputs into electrical signals controlling the aircraft's control surfaces.*

Orthogonal Analytical Perspectives:

1. Aeronautical Engineering Perspective:

- *Focuses on aerodynamic principles and flight physics*
- *Analyzes control surface response characteristics*
- *Evaluates flight envelope protection mechanisms*
- *Question asked: "Does the system maintain proper aerodynamic control under all flight conditions?"*

2. Software Engineering Perspective:

- *Examines software architecture and control algorithms*
- *Analyzes code reliability and error handling*
- *Evaluates software verification and validation approaches*
- *Question asked: "Is the software robust against failures and unexpected inputs?"*

3. Electrical Engineering Perspective:

- *Studies power distribution and redundancy*
- *Analyzes signal integrity and electromagnetic interference*
- *Evaluates hardware component reliability*
- *Question asked: "Will the electrical systems function reliably in all environmental conditions?"*

4. Human Factors Perspective:

- *Examines pilot interface design and feedback mechanisms*
- *Analyzes workload and situational awareness factors*
- *Evaluates potential for mode confusion*
- *Question asked: "Can pilots effectively interact with and monitor the system?"*

Integration Point: *A critical safety issue is identified only when these perspectives are combined: in certain rare electronic interference conditions, the system might generate subtle control anomalies that fall within software parameter limits and electrical specifications, but create handling characteristics that pilots find difficult to recognize and compensate for during high-workload situations.*

This example demonstrates how orthogonal views can identify complex interaction issues that no single disciplinary analysis would discover on its own.

6.3.3 Visualization and Coordination

Orthogonal perspectives are represented as independent analytical "vectors." Intersections among these vectors reveal overlooked issues or redundant analyses, allowing targeted, prioritized risk mitigation. Effective interdisciplinary collaboration is needed to get the most from orthogonal analyses.

6.3.4 Benefits and Challenges

6.3.4.1 Benefits

- **Comprehensive Coverage:** Minimizes analytical blind spots through diverse viewpoints
- **Confirmation Through Convergence:** When independent perspectives reach similar conclusions, confidence increases

- **Identification of Interactions:** Reveals system behaviors that emerge from component interactions
- **Bias Reduction:** Mitigates individual analyst biases through multiple perspectives
- **Stakeholder Integration:** Incorporates diverse stakeholder concerns into analysis

6.3.4.2 Challenges

- **Communication Overhead:** Requires effective communication across disciplinary boundaries
- **Expertise Availability:** Demands experts from multiple domains
- **Integration Complexity:** Can be difficult to systematically combine findings from different perspectives
- **Conflicting Priorities:** Different domains may have competing objectives or priorities
- **Resource Intensity:** Requires more time and resources than single-perspective analysis

The point of orthogonal views is not to create more meetings. It is to let different specialties overlap just enough to find surprises early. When you finish an analysis, write one or two sentences on what each perspective added; that short reflection often pays for itself on the next project.

6.4 Applying Analytical Methods

Example 6.4: ***From Analysis to Design Action.*** *An FMEA identifies "loss of wheel speed input" with severity II and high detectability. The resulting action is to add sensor plausibility checks (cross-compare left/right axles) and a watchdog on stale data. The analytical finding drives a concrete change: a safe fallback that limits traction/braking effort and raises a clear diagnostic, with test cases linked to the FMEA row.*

Integrating deterministic, probabilistic, and orthogonal approaches gives you thorough risk assessment capabilities.

6.4.1 Implementation Steps

1. **Problem Definition:** Clearly establish analytical objectives, scope, and required outputs.
2. **Method Selection:** Choose analytical methods aligned with problem specifics—deterministic for explicit logic verification, probabilistic for

uncertainty handling, and orthogonal for comprehensive multidisciplinary evaluation.

3. **Analysis Execution:** Perform structured analyses using methodologies like Fault Tree Analysis, FMEA, and reliability modeling.
4. **Interpretation:** Evaluate results critically, identifying significant risks, critical components, and systemic vulnerabilities.
5. **Mitigation Planning:** Recommend risk reduction measures such as redundancy, improved testing procedures, preventive maintenance, or design modifications based on analysis findings.
6. **Validation of Analysis:** Verify analytical results through independent review, testing, or operational feedback.
7. **Documentation and Communication:** Create clear, comprehensive records of analysis methods, assumptions, results, and limitations.

6.4.2 Methodical Integration Framework

A good framework for integrating multiple analytical approaches involves:

- **Sequential Application:** Begin with deterministic analysis to verify basic system functionality and safety features, then apply probabilistic methods to address uncertainty, followed by orthogonal analysis to ensure comprehensive coverage.
- **Iterative Refinement:** Use findings from each method to refine and focus on subsequent analyses, creating a progressively more detailed and accurate risk assessment.
- **Cross-validation:** Compare results from different methods to identify consistencies and discrepancies, providing higher confidence in areas of agreement and focusing attention on areas of disagreement.
- **Combined Risk Profiles:** Develop integrated risk assessments that incorporate deterministic compliance status, probabilistic risk metrics, and insights from multiple domain perspectives.

6.4.3 Detailed Example Application

Automated Braking System in Autonomous Vehicles:

- **Deterministic Analysis:** Verifies braking logic accuracy under defined scenarios, including standard operation and known failure modes.
- **Probabilistic Analysis:** Estimates failure probabilities for braking sensors and actuators, using operational performance data and environmental variables.

- **Orthogonal Analysis:** Combines automotive safety expertise, system engineering insights, and regulatory compliance perspectives to produce a complete risk profile.
- **Outcomes and Recommendations:** Analysts identify critical risk scenarios, guiding decisions to incorporate redundant braking systems, advanced diagnostic mechanisms, and rigorous maintenance and inspection protocols.

6.5 Advanced Analytical Techniques

Beyond the fundamental methods, several advanced analytical techniques provide specialized capabilities for safety-critical systems.

6.5.1 Model-based Safety Analysis

This approach integrates safety analysis directly with system models, enabling:

- Automated generation of failure analyses from system models
- Real-time safety assessment as designs evolve
- Direct traceability between design elements and safety properties
- Simulation-based validation of safety mechanisms

6.5.2 Bayesian Methods

Bayesian statistical approaches provide powerful tools for:

- Incorporating prior knowledge with new evidence
- Updating risk assessments as operational data accumulates
- Handling sparse or incomplete data sets
- Quantifying uncertainty in risk estimates

6.5.3 System Theoretic Process Analysis

This holistic approach focuses on system-level properties rather than component failures:

- Analyzing safety as an emergent property of system interactions
- Identifying unsafe control actions and scenarios
- Evaluating feedback and control loops for safety implications
- Addressing complex socio-technical system aspects

Key Takeaway

- Analytical methods are key tools for evaluating and improving safety-critical system reliability.

Part IV

Perspectives

7

Common Fallacies

After reading this chapter, you should be able to:

- Spot a few traps in real text (e.g., "safer than safe," standard worship).
- Show the signal that gave it away and a small counter-move.
- Run a short five-whys on a safety miss and record the fix.

Designing, analyzing, and implementing safety-critical systems involves navigating numerous pitfalls. Misconceptions, oversimplifications, and flawed reasoning frequently compromise safety, creating preventable risks. This chapter identifies common fallacies encountered in safety-critical systems and explains them through practical examples.

7.1 Independence of Events

For the formal treatment of independence, see Section 2.1.

Example 7.1: ***False Independence.*** *Two wheel-speed sensors are placed on opposite axles and treated as independent. A shared harness routing near a heat source causes intermittent opens on both lines after prolonged operation. The correlation invalidates the independence assumption. Fix: redesign routing and add cross-channel plausibility checks with environmental stress screening.*

One of the most pervasive fallacies in safety engineering is the assumption that failures in redundant components occur independently without rigorous verification. Engineers often calculate vanishingly small failure probabilities by multiplying individual component failure rates, but this arithmetic only holds when components truly have no common vulnerabilities. See Figure 7.1 for a high-level pattern of how these misconceptions reinforce each other.

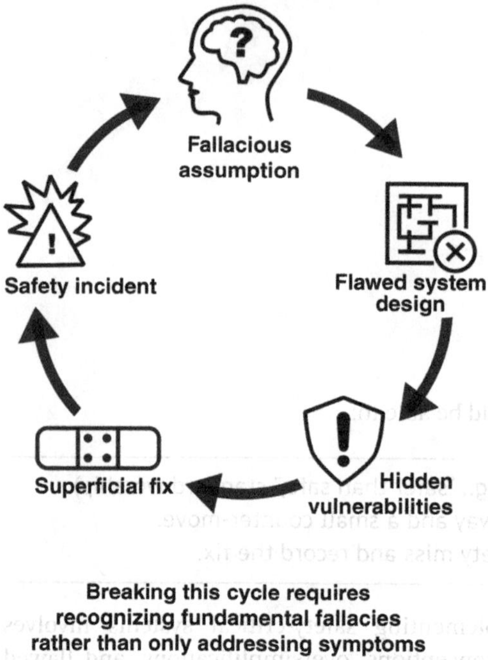

Figure 7.1 The safety fallacy cycle.

Consider the statement: "If one sensor fails, the others will continue to function normally." This reasoning appears sound on the surface, but reality tells a different story. Common causes such as environmental conditions, power disruptions, or software glitches can simultaneously impact multiple sensors, invalidating the independence assumption entirely.

The danger of overestimating independence lies in the false confidence it creates. Design teams may specify inadequate redundancy or develop insufficient hazard management strategies because they believe the calculated probability of simultaneous failure is negligibly small. When common-cause failures actually occur, the system has no remaining defense.

To guard against this fallacy, teams should conduct rigorous common-cause failure analyses (CCFA), perform dependency mapping, and engage in structured redundancy planning that actively seeks out hidden dependencies. The goal is not merely to assume independence but to design it into the system through deliberate diversity of components, suppliers, power sources, and environmental exposures.

Example 7.2: *Consider a modern aircraft with three redundant air data systems. Each system includes sensors measuring airspeed, altitude, and temperature.*

Fallacious Thinking: *The system designer assumes these three systems fail independently, calculating a negligible probability of simultaneous failure (e.g., 10^{-9} probability if each has a 10^{-3} failure rate).*

Reality: *All three systems might:*

- *Share a common firmware component with a subtle bug that activates under specific flight conditions*
- *Be affected by the same icing conditions when flying through certain cloud formations*
- *Process data using similar algorithms vulnerable to the same numerical instability*
- *Receive power from buses with a shared connection point*

Proper Approach: *The designer should:*

- *Identify all potential common causes (physical placement, power sources, environmental factors, software components)*
- *Intentionally diversify designs (different sensor technologies, separate power sources, varied firmware implementations)*
- *Test specifically for common-mode failures (simultaneous exposure to extreme conditions)*
- *Quantify the actual level of independence through analysis and testing*

This example demonstrates how assumed independence can vastly underestimate actual failure probabilities, potentially by several orders of magnitude.

7.2 Complexity Versus Probability

A seductive fallacy holds that well-engineered, complex systems inherently reduce failure likelihood. The intuition seems reasonable: more lines of code means more functionality, more checks, and presumably more safety. Figure 7.2 sketches why increasing complexity does not necessarily reduce event probability—and can raise correlated risks.

Consider the assertion: "The software comprises thousands of thoroughly tested lines of code, hence it must be highly reliable." This logic inverts the actual relationship between complexity and reliability. In practice, increased complexity creates unforeseen interactions between components, complicates failure mode identification, and introduces hidden vulnerabilities that may not manifest until specific operating conditions arise.

The danger of this fallacy lies in how complexity masks latent risks. Each additional component, interface, and code path represents another potential failure

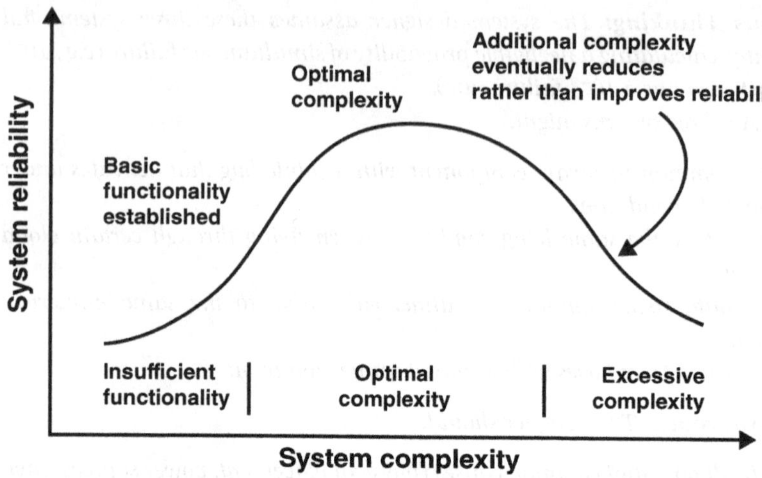

Figure 7.2 Complexity vs. reliability relationship.

point. The more complex the system, the harder it becomes to anticipate all possible failure combinations, increasing the probability of critical failures that escape testing and analysis.

The remedy is to embrace simplicity as a design virtue. Teams should adopt modular, simplified system designs wherever possible, enforce rigorous interface management to contain the effects of component failures, and prioritize testing of critical functional areas. When complexity is unavoidable, it should be carefully isolated and subjected to proportionally more intensive verification.

7.3 Safer Than Safe

Example 7.3: ***"Safer than Safe."*** *A protective threshold is tightened to reduce a hazard, but the new threshold triggers frequent nuisance trips that operators learn to override. The overall risk increases. Fix: tune thresholds against operational data, add hysteresis, and monitor override rates as a leading indicator.*

The assumption that exceeding established safety standards automatically enhances system safety reflects a fundamental misunderstanding of how safety works in practice. Safety engineering is about achieving an appropriate balance, not about maximizing every protective measure without regard for consequences.

Consider the logic: "Doubling the time required for emergency system activation will automatically increase safety." On the surface, this seems sensible—more margin should mean more protection. But excessive safety features can produce counterproductive effects. Operators who encounter frequent false alarms develop complacency and learn to dismiss or override warnings. Interfaces burdened with excessive confirmations slow critical responses. Safety interlocks that trigger too easily get bypassed. The net result can be reduced effectiveness and new usability issues that undermine the original safety intent.

The hazard of "safer than safe" thinking is that overly cautious measures may inadvertently create new hazards or negatively impact operational performance in ways that degrade overall safety. A system that operators learn to work around is less safe than one they trust and use as designed.

The appropriate response is to customize safety measures for specific operational contexts. This requires empirical testing with realistic scenarios, simulations that capture actual operating conditions, and stakeholder feedback from the people who will use the system daily. The goal is safety measures that operators respect and follow, not ones they learn to circumvent.

Example 7.4: *Consider a pharmaceutical manufacturing process that requires sterile conditions.*

Fallacious Thinking: *The process engineer decides that if the standard sterilization procedure (heating equipment to 121 °C for 15 minutes) is effective, then extending the sterilization to 45 minutes must be three times safer.*

Reality: *This excessive approach creates several problems:*

- *Extended heating damages equipment seals, creating potential contamination pathways*
- *Longer sterilization cycles reduce production capacity, pressuring operators to take shortcuts elsewhere in the process*
- *The additional time increases energy usage without providing measurable safety benefits*
- *Staff become complacent about other critical sterilization parameters (like pressure or temperature verification)*

Proper Approach: *The engineer should:*

- *Validate the standard sterilization procedure against specific contaminants of concern*
- *Identify the critical parameters that actually affect sterilization effectiveness*
- *Implement monitoring and verification of those parameters rather than arbitrarily extending times*
- *Focus improvement efforts on process consistency and staff training*

This example demonstrates how "more is better" thinking can actually introduce new failure modes while creating a false sense of security.

7.4 Safety Through Testing

Example 7.5: ***Testing Guarantees Safety.*** *A subsystem with rare timing faults passes extensive unit tests but fails under integrated load due to scheduler interactions. Fix: add stress/scenario testing, time-budget analysis, and interface-level assertions. Testing is evidence, not a guarantee.*

Testing is essential to safety assurance, but the belief that extensive testing alone guarantees system safety represents a dangerous oversimplification. Testing can only reveal the presence of defects in the scenarios actually executed; it cannot prove the absence of defects in scenarios not tested.

The claim "We have executed thousands of test cases without errors, hence the system is completely safe" illustrates this fallacy clearly. While passing thousands of tests provides valuable evidence, it tells us nothing about the system's behavior in scenarios those tests did not cover. Testing uncovers existing defects but cannot guarantee the absence of undiscovered flaws, particularly those that emerge only in edge cases or under exceptional conditions that were never anticipated or tested.

The danger of overreliance on testing alone is that it creates false confidence while neglecting latent or complex failure modes. Systems that pass all tests may still harbor defects that activate under conditions the test suite never explored—timing races that occur only under heavy load, numerical instabilities that appear only at specific input ranges, or interaction effects between components that were tested in isolation.

Robust safety assurance requires complementing testing with other verification methods. Formal verification can mathematically prove properties hold across all inputs. Fault injection deliberately stresses the system to reveal hidden weaknesses. Robust hazard analyses identify what could go wrong before testing begins. Simulations of rare but plausible scenarios explore operating conditions that would be impractical or dangerous to test with real hardware. Together, these approaches provide the defense in depth that testing alone cannot achieve.

7.5 Independence in the Design Process

Independence between design and verification functions is a cornerstone of safety engineering, but the belief that separation of design and verification teams automatically ensures improved safety outcomes mistakes organizational structure for actual effectiveness.

The reasoning "Separating design and test teams ensures objective and accurate evaluations" captures only part of the picture. Independence provides objectivity—verifiers who did not create the design are more likely to question assumptions and find overlooked defects. However, independence without effective communication creates its own problems. Independent teams may develop knowledge gaps about design intent, operational context, or implicit assumptions. These gaps lead to misunderstandings, overlooked issues, and verification activities that fail to address the most critical risks.

The danger materializes when independent teams operate in isolation. Verifiers may test against requirements without understanding why those requirements exist, missing hazards the requirements failed to capture. Designers may dismiss verification findings as reflecting ignorance of constraints they never communicated. The result is miscommunication, missed defects, and inefficiencies that consume resources without improving safety.

The solution is not to abandon independence but to structure it properly. Effective independence requires structured collaboration processes that maintain objectivity while ensuring knowledge transfer. Clear communication protocols should define when and how teams share information. Shared documentation makes design rationale and verification findings accessible to all parties. Accountability mechanisms between design and verification teams ensure that identified issues receive appropriate attention and resolution.

7.6 Excessive Maintenance Is Not a Safety Issue

The intuition that more maintenance equals more safety leads many organizations to assume that frequent maintenance inherently guarantees safety. This reasoning overlooks a critical insight: every maintenance intervention is itself an opportunity for something to go wrong.

Consider the belief: "Regularly replacing sensors ensures continuous, optimal system performance." While regular maintenance has its place, excessive maintenance activities introduce their own risks. Each time a technician opens a panel, disconnects a cable, or handles a component, there is potential for human error. Connectors may not be fully reseated. Calibrations may be performed incorrectly. Adjacent components may be inadvertently disturbed. The system may function worse after maintenance than before, introducing unintended system faults.

The danger of this fallacy is that unnecessary maintenance tasks increase the likelihood of introducing new hazards or errors. Organizations that mandate maintenance intervals without regard for actual component reliability may be creating more risk than they prevent. Maintenance fatigue can also set in: technicians performing the same procedure for the hundredth time may rush through steps or skip checks that seem redundant.

A more effective approach emphasizes designing systems for reliability to reduce maintenance demands in the first place. When maintenance is necessary, provide comprehensive training that explains not just procedures but the reasoning behind each step. Clearly documented maintenance procedures reduce variability and help manage risks effectively. Condition-based maintenance—triggered by actual component degradation rather than arbitrary schedules—often provides better safety outcomes than calendar-based replacement policies.

Example 7.6: *Consider a critical industrial control system that measures and regulates chemical flow rates.*

Fallacious Thinking: *The maintenance manager implements a policy requiring monthly calibration of all flow sensors, regardless of their performance history, believing this will maximize system safety.*

Reality: *This excessive maintenance approach creates several problems:*

- *Each calibration requires system shutdown and sensor handling, introducing opportunities for human error*
- *Frequent disconnection of sensors damages connectors and seals over time*
- *Technicians forced to perform unnecessary calibrations become complacent and may rush procedures*
- *Some sensors actually perform worse immediately after recalibration due to settling effects*
- *The frequent interventions interrupt production and create pressure to rush maintenance procedures*

Proper Approach: *The manager should:*

- *Implement condition-based maintenance using sensor performance data*
- *Establish drift thresholds that trigger calibration only when needed*
- *Track calibration history to identify sensors requiring replacement rather than perpetual adjustment*
- *Design maintenance procedures that minimize system disruption and human error*

This example demonstrates how excessive maintenance can actually decrease system reliability by introducing more opportunities for error than it prevents.

7.7 Complexity Is Not a Safety Issue

Some organizations treat complexity as orthogonal to safety, overlooking the safety implications inherent to system complexity. This compartmentalized thinking leads to designs that grow increasingly intricate without corresponding increases in safety analysis effort.

The assertion "System complexity does not inherently impact our safety management approach" reflects this misconception. In practice, increased complexity directly obscures potential failure modes, making risk identification and management progressively more difficult. Every added component, interface, and operational mode expands the space of possible system behaviors, including failure behaviors that analysts must anticipate and mitigate.

Complex systems are particularly prone to hidden, cascading failures—chains of events where one malfunction triggers another in ways that are difficult to predict or control. The interactions between components may be too numerous to analyze exhaustively, and subtle timing or sequencing dependencies may only manifest under specific conditions. When failures do occur, the complexity that created them also makes diagnosis and recovery more challenging.

Addressing this fallacy requires treating complexity as a first-class safety concern. Teams should prioritize simplicity in system design, resisting the temptation to add features or capabilities that increase complexity without commensurate safety analysis. Regular complexity analyses should assess whether the system remains tractable for safety assurance. Fault-tolerant architectures help contain the effects of failures that complexity makes inevitable. Structured hazard identification methodologies provide systematic approaches to exploring failure possibilities even when the system is too complex for intuitive analysis.

7.8 Latent Failures

A particularly insidious fallacy involves underestimating risks posed by hidden, latent system failures. These are defects or degradations that exist within the system but have not yet manifested as observable malfunctions. Figure 7.3 illustrates a typical progression from hidden defect to incident.

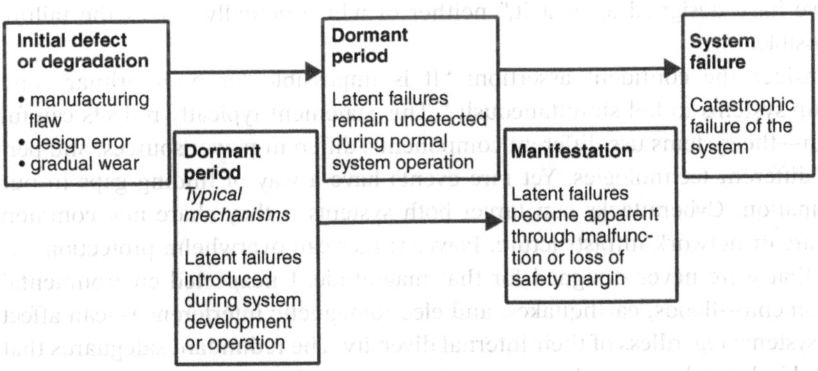

Figure 7.3 The latent failure pathway.

The reasoning "System stability to date implies latent failures are improbable" confuses the absence of symptoms with the absence of disease. Latent issues like calibration drift, sensor degradation, or software aging can remain hidden for extended periods, accumulating undetected until some triggering condition causes them to manifest as severe system failures. A sensor that has gradually drifted out of tolerance may continue to provide plausible readings until conditions push it beyond a threshold where its errors become safety-critical.

The danger of latent failures lies precisely in their invisibility. Unaddressed latent failures can escalate unnoticed into major incidents, often at the worst possible moment—when the system is under stress and its safety margins are already reduced. Because these failures develop gradually, they may defeat protective mechanisms designed to detect sudden changes.

Effective defense against latent failures requires proactive detection rather than passive observation. Organizations should implement periodic diagnostics that actively probe system components for degradation, even when those components appear to be functioning normally. Predictive maintenance techniques use trend analysis to identify components approaching failure before they actually fail. Continuous system health monitoring tracks key parameters over time, providing early warning of drift or degradation that could eventually compromise safety. The goal is to identify latent failures while they are still latent, before they manifest as operational incidents.

7.9 Impossible but Conceivable Failure Mode

Engineers sometimes dismiss unlikely but plausible system failures as "impossible," effectively removing them from consideration in hazard analyses. This fallacy substitutes linguistic convenience for rigorous probability assessment—"impossible" really means "I cannot imagine how it would happen" or "we have designed against it," neither of which actually makes the failure impossible.

Consider the confident assertion: "It is impossible for both primary and backup systems to fail simultaneously." This statement typically reflects careful design—the systems use different components, different power sources, and perhaps different technologies. Yet rare events have a way of finding gaps in our imagination. Cyberattacks can target both systems if they share any common software or network infrastructure. Power surges can overwhelm protection circuits that were never designed for that magnitude. Unexpected environmental phenomena—floods, earthquakes, and electromagnetic interference—can affect both systems regardless of their internal diversity. The redundant safeguards that seemed independent may share vulnerabilities their designers never anticipated.

Ignoring low-probability, high-impact scenarios exposes systems to severe vulnerabilities. The scenarios that cause the worst accidents are often precisely those that were dismissed as "impossible"—nobody planned for them, nobody tested for them, and nobody developed recovery procedures for them.

Robust safety engineering includes extreme yet plausible scenarios in hazard analyses, specifically exploring what could go wrong if multiple safeguards failed simultaneously. Design strategies should employ diverse redundancy that truly eliminates common-cause failure paths, not just the obvious ones. Risk management plans should address worst-case conditions even when those conditions seem remote, because the consequences of being wrong are severe.

Example 7.7: *Consider a hospital emergency backup power system with dual generators that operate independently.*

Fallacious Thinking: *The hospital administrator declares, "It's impossible for both generators to fail simultaneously—we've ensured they're completely independent systems with separate fuel supplies, starting mechanisms, and maintenance schedules."*

Reality: *Despite apparent independence, several scenarios could cause simultaneous failure:*

- *Both generators use the same model of electronic control module with an undiscovered firmware defect*
- *A major flood affects both generator rooms despite being in different building wings*
- *Both fuel supplies become contaminated by the same batch of poor-quality fuel from a single supplier*
- *A coordinated cyberattack targets both generator control systems*
- *An electromagnetic pulse affects all electronic systems regardless of their physical separation*

Proper Approach: *The administrator should:*

- *Explicitly identify and analyze potential common failure modes*
- *Implement true diversity (different generator technologies from different manufacturers)*
- *Develop emergency procedures assuming both generators could fail*
- *Consider tertiary backup systems (battery banks) with fundamentally different technologies*
- *Test the entire backup system under realistic stress conditions*

This example illustrates how declaring certain failures "impossible" often reflects limited imagination rather than true system resilience.

7.10 Personnel Are Infallible

A persistent fallacy in safety engineering is assuming that trained personnel will always operate flawlessly. This belief overestimates what training can achieve and underestimates the inherent variability of human performance, particularly under stress.

The expectation "Trained operators will never deviate from emergency protocols" reflects this misconception. Training certainly improves performance and consistency, but it cannot eliminate human error. Human errors arise from fundamental cognitive limitations that no amount of training can overcome. Stress narrows attention and impairs judgment. Fatigue degrades vigilance and reaction time. Unclear instructions create opportunities for misinterpretation. Time pressure forces shortcuts. These factors are persistent risks in any human-operated system, and they intensify precisely during emergencies when flawless performance matters most.

Overestimating human reliability is dangerous because it leaves critical systems vulnerable to predictable human limitations during high-stress operations. Systems designed with the assumption that operators will always perform perfectly provide no defense when operators inevitably do not. The result is that the first line of defense fails exactly when conditions are most demanding.

Effective safety design acknowledges human fallibility and compensates for it. Organizations should automate safety-critical functions where possible, removing humans from loops where their errors could have severe consequences. User interfaces should be intuitive, minimizing the cognitive load on operators and reducing opportunities for confusion. Regular realistic training helps operators maintain skills and builds muscle memory for emergency responses. Clearly documented emergency procedures provide a reference when stress impairs memory and judgment. The goal is a system that remains safe despite occasional human errors, not one that depends on perfection.

7.11 Normalization of Deviance

One of the most dangerous organizational fallacies involves gradually accepting increasingly risky practices as "normal" when they don't immediately result in failures. This phenomenon, identified in studies of major accidents, including the space shuttle disasters, represents a systematic erosion of safety culture that occurs incrementally, almost invisibly.

The reasoning is seductive: "We've been bypassing this safety check for months without incident, so it must not be necessary." Each successful deviation from standard procedure reinforces the belief that the deviation is acceptable.

The absence of immediate negative consequences is interpreted as evidence that the safety margin was excessive, leading to gradual erosion of safety margins and acceptance of increasingly risky behaviors.

This process is particularly insidious because it normalizes dangerous practices through a series of small, seemingly reasonable steps. Each deviation is only slightly more risky than the last. Each decision seems justified by the absence of problems so far. The organization drifts toward disaster without anyone making a conscious decision to accept unacceptable risk. When an accident finally occurs, investigators often find a long history of warning signs that were systematically rationalized away.

Combating normalization of deviance requires deliberate organizational effort. Strict adherence to safety protocols must be maintained even when deviations seem harmless. Regular safety culture assessments should examine whether actual practices match documented procedures. Anonymous reporting systems enable personnel to raise concerns without fear of retaliation. Trend analysis of near-misses and procedure deviations can identify creeping deviations before they culminate in an incident. The key insight is that the absence of accidents is not evidence of safety—it may simply mean the organization has not yet encountered the conditions that will expose its accumulated vulnerabilities.

Example 7.8: *Consider a manufacturing facility with a lockout-tagout procedure for machinery maintenance.*

Initial Procedure: *Full power shutdown, verification of zero-energy state, physical locks applied by each worker, documentation signed.*

Gradual Deviation:

- *Week 1: A maintenance task runs late, and workers skip the documentation step "just this once" to finish on time. No incident occurred.*
- *Month 1: For "quick fixes" lasting under 5 minutes, workers begin only using the emergency stop button rather than full power disconnection. No incidents occurred.*
- *Month 3: Workers start performing "simple adjustments" without any lockout procedure when production deadlines are tight. Still no incidents.*
- *Month 6: The partial shutdown becomes standard practice for most maintenance tasks since "we've never had a problem."*
- *Month 8: A machine unexpectedly reactivates during maintenance, causing a severe injury.*

Root Issue: *Each deviation that didn't result in immediate consequences reinforced the belief that the safety procedure was unnecessarily cautious. The absence of negative feedback created a false sense of security, while each accepted deviation made the next one seem more reasonable.*

Proper Approach: *Management should:*

- *Regularly audit actual practices against documented procedures*
- *Investigate and address the reasons why workers feel pressured to take shortcuts*
- *Implement a no-blame reporting system for safety procedure violations*
- *Share near-miss stories from other facilities to make potential consequences more tangible*

This example demonstrates how seemingly reasonable incremental deviations can accumulate into dangerous practices that become normalized within an organization.

7.12 System Safety Is Just Component Safety

A fundamental misconception in safety engineering is assuming that ensuring the safety of individual components automatically guarantees overall system safety. This reductionist view treats safety as a property that can be assembled from parts, like building blocks, rather than recognizing it as something that emerges from the relationships between parts.

The claim "All our components meet their safety requirements, therefore the system as a whole is safe" reflects this misunderstanding. Each component may indeed function safely within its specified operating envelope. But safety is an emergent property of the entire system, arising from how components work together rather than from the components themselves. Many hazards originate not from component failures but from component interactions, integration issues, and system-level behaviors that are simply not visible when examining components in isolation.

This reductionist thinking is dangerous because it systematically misses interaction effects, boundary conditions, and emergent behaviors that often cause the most severe accidents. A sensor that meets all its requirements may still contribute to an accident if the system misinterprets its outputs under conditions its designers never anticipated. Two components that function flawlessly in isolation may interfere with each other when integrated. The timing relationships between components may create hazardous race conditions that no component-level analysis would reveal.

Addressing this fallacy requires shifting perspective from components to systems. Teams should implement system-level safety analyses that explicitly examine how components interact and how hazards emerge from those interactions. Thorough integration testing exercises the system as a whole, not just its parts. Simulation of full-system behavior explores operating conditions that would be impractical to test physically. Systematic analysis of component interactions under various conditions reveals hazards that component specifications alone cannot capture.

Key Takeaways

- Recognizing and actively mitigating common fallacies in safety-critical systems is needed for robust and reliable designs.
- By systematically questioning assumptions, applying rigorous analytical methods, and continuously improving processes, organizations improve overall safety, reliability, and effectiveness.
- The fallacies examined in this chapter share common characteristics:
  - They stem from cognitive biases that affect even experienced engineers
  - They often appear reasonable on the surface
  - They become more dangerous when unquestioned
  - They persist because their consequences may not be immediately apparent
  - They can be countered through structured analysis and healthy skepticism
- To counter these fallacies, safety practitioners should:
  - Cultivate a questioning attitude that challenges assumptions
  - Apply diverse analytical techniques
  - Seek independent reviews of safety-critical designs
  - Learn from historical failures
  - Build organizational cultures that reward identifying potential issues
- Proactive and deliberate safety management must remain central to safety-critical system design.

8

Dirty Little Secrets

After reading this chapter, you should be able to:

- Map how schedule or politics can bend a safety call in your project.
- Set a clear escalation path for "this is not safe."
- Write a short rule for who signs what, and with what proof.

Safety-critical systems operate within environments characterized by high expectations, rigorous requirements, and substantial costs. Yet there are certain uncomfortable truths about their design, implementation, and operation that are rarely openly discussed. These "dirty little secrets" reveal complexities, difficult trade-offs, and human factors that affect safety-critical systems.

8.1 Bug-free Software

Example 8.1: ***Zero Defects ≠ Safe.*** *A subsystem achieves 100% unit-test pass rate and zero known defects at release, yet a hazard remains because the design never constrained an unsafe operating mode under a credible sensor failure. Fix: trace tests to hazards and safety requirements, and include fault-injection scenarios that exercise unsafe modes, not just nominal behavior.*

The Myth: Software in safety-critical systems must be entirely bug-free. See Figure 8.1 for a high-level map of myths and realities discussed in this chapter.
The Reality: Completely bug-free software is practically unattainable. Despite rigorous methods like formal verification, static analysis, and exhaustive testing, subtle errors remain inevitable. Safety-critical systems emphasize managing and mitigating critical bugs rather than striving to eliminate every possible bug. Good practices include clear software segmentation, well-defined interfaces, plausibility checks on data, and mechanisms that contain and limit the impact of any residual errors.

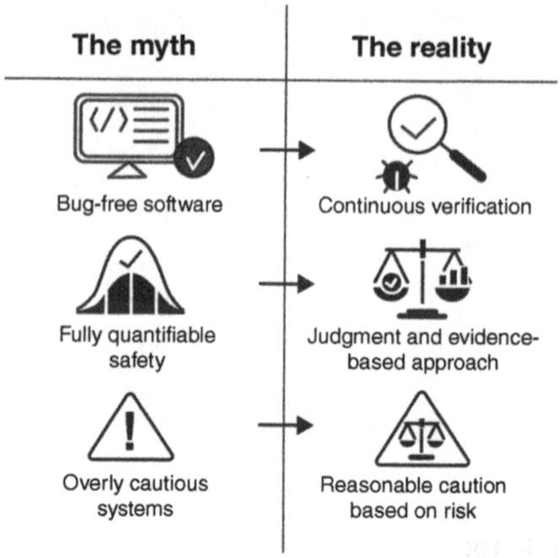

Figure 8.1 The safety myth vs. reality framework.

The aviation industry exemplifies this approach through design assurance levels (DAL) and partitioned architectures where lower-criticality functions cannot interfere with higher-criticality functions. Medical devices similarly employ watchdog timers and fault containment regions to prevent error propagation.

Example 8.2: *Consider a radiation therapy system that delivers precise doses of radiation to cancer patients:*

Myth-based Approach: *Developers attempt to create perfect, bug-free control software through extensive testing alone, assuming all bugs can be eliminated.*

Reality-based Approach: *Developers implement multiple layers of protection:*

- ***Primary Control System:*** *Main dosage calculation and delivery software undergoes rigorous verification but is assumed to potentially contain unknown bugs*
- ***Independent Verification System:*** *Separate, simpler software running on different hardware independently calculates expected dosage ranges*
- ***Hardware Interlocks:*** *Physical mechanisms prevent radiation delivery exceeding maximum safe limits, regardless of software commands*
- ***Monitoring System:*** *Real-time comparison between intended and actual delivery with automatic shutdown if discrepancies exceed thresholds*
- ***Operational Procedures:*** *Treatment plans reviewed by multiple clinicians before treatment begins*

This approach acknowledges the inevitability of bugs while creating multiple, diverse barriers to prevent any single bug from causing harm. Rather than pursuing the myth of perfect software, the system is designed to remain safe despite potential software flaws.

Why It's Uncomfortable: Admitting bugs exist challenges perceptions of absolute safety, compelling engineers to add protective mechanisms that are subjective, computationally intensive, and difficult to comprehensively validate. This acknowledgment also raises difficult questions about legal liability and public perception.
Takeaway: Design systems with resilience, redundancy, and defensive programming. Focus efforts on minimizing the likelihood and impact of critical issues by thoroughly analyzing data flows and potential error propagation paths. Implement mechanisms like runtime monitoring, graceful degradation modes, and independent safety channels to manage the inevitable presence of software defects.

8.2 Quantification of Safety

For the formal framing, see Section 1.3.1.
The Myth: Safety can be precisely quantified. Figure 8.2 summarizes how risk categories emerge from severity and likelihood, and why quantification is bounded by assumptions.

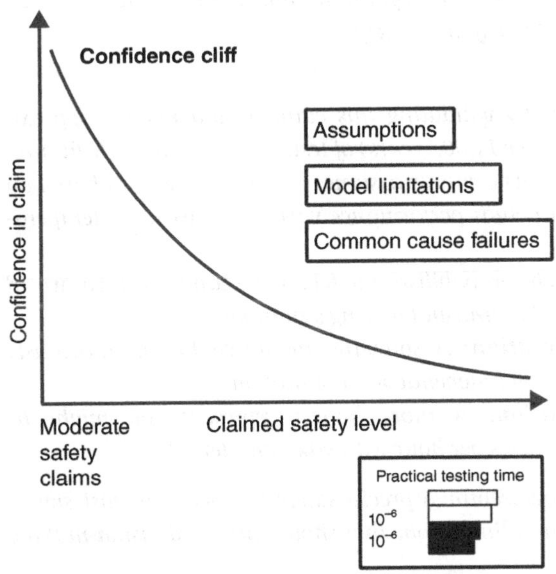

Figure 8.2 The confidence paradox in safety quantification.

The Reality: Safety quantification relies heavily on assumptions, incomplete data, and probabilistic projections that are inherently uncertain. Small variations in underlying assumptions or reliability data can dramatically affect safety metrics. Industries like nuclear power, aviation, and medical devices use techniques such as fault tree analysis (FTA), failure mode and effects analysis (FMEA), and probabilistic risk assessment (PRA), but these methods depend critically on input quality and model fidelity.

Common-cause failures, systematic errors, and rare but consequential "black swan" events further complicate quantification, often escaping traditional analysis methods. The statistical validation of extremely low failure rates (e.g., 10^{-9} failures per hour) would require impractically long testing periods—sometimes exceeding the system's operational lifetime.

Why It's Uncomfortable: Quantitative safety metrics provide an illusion of certainty and can lead to false confidence. The scales involved (e.g., hazard rates such as 10^{-9} per hour) are difficult to conceptualize, challenging intuitive understanding. Decision-makers may overlook the uncertainty bounds and model limitations when interpreting these metrics.

Example 8.3: *Consider an autonomous vehicle's collision avoidance system with a claimed failure rate of 10^{-8} per hour of operation.*

Myth-based Approach: *Stakeholders accept this number at face value, assuming it represents an objective truth about system safety.*

Reality-based Analysis:

- ***Testing Limitation:*** *Directly validating this claim would require approximately 100 million hours (over 11,000 years) of testing without a single failure*
- ***Assumption Sensitivity:*** *The analysis assumes sensor failure rates based on laboratory conditions, but actual performance varies with weather, temperature, and aging*
- ***Model Incompleteness:*** *The probabilistic model doesn't capture all real-world scenarios, particularly novel or unusual driving situations*
- ***Common-cause Consideration:*** *A software update could simultaneously affect all vehicles, invalidating independence assumptions*
- ***Uncertainty Representation:*** *A more honest representation might be "between 10^{-7} and 10^{-9} failures per hour with 90% confidence"*

This example demonstrates how seemingly precise safety numbers can mask significant underlying uncertainties and limitations that should inform decision-making and risk management.

Takeaway: Use quantitative safety metrics as guidance rather than absolute truths. Always supplement these metrics with qualitative insights, realistic margins, and robust sensitivity analyses. Carefully document assumptions and regularly reassess them to avoid unjustified confidence in numerical outcomes. Recognize that safety is an emergent property that cannot be fully captured through numerical analysis alone.

8.3 Documentation Equals Safety

Example 8.4: ***Paper Safety.*** *A project maintains impeccable documentation and review checklists while an early assumption (track friction coefficient) is never validated. The system meets every process checkpoint but violates a safety margin in winter operations. Fix: require evidence types per claim (test/analysis/inspection), surface unverified assumptions, and link them to targeted validation tasks.*

The Myth: Extensive documentation ensures system safety.
The Reality: While essential for clarity and communication, extensive documentation alone does not guarantee safety. Poor documentation—outdated, overly complex, or unclear—can undermine rather than enhance system understanding and safety assurance. Documentation must be actively maintained, accessible, and actually used by development teams to have a real impact.

In practice, safety-critical documentation often suffers from several problems: inconsistencies between artifacts, circular references, excessive detail obscuring critical information, and disconnection from actual implementation. The aviation industry's experience with aircraft maintenance documentation demonstrates how documentation complexity can inadvertently contribute to maintenance errors rather than prevent them.
Why It's Uncomfortable: Organizations may prioritize documentation merely to satisfy auditors or regulators, creating "paper safety" rather than genuine robustness. This compliance-driven approach can drain resources while giving a false sense of security. Engineers may become disillusioned when documentation efforts don't translate to tangible safety improvements.
Takeaway: Aim for concise, clear, and purposeful documentation. Ensure documentation effectively communicates essential safety information and helps understanding rather than obscuring key insights. Avoid analyses and documentation that become larger or more complex than the system itself. Focus on traceability between requirements, design decisions, and verification evidence to create living documentation that genuinely supports safety.

8.4 Process Can Be "Backfilled"

The Myth: Safety processes can be retroactively applied to existing designs.
The Reality: Retroactively applying a robust safety process to an existing design is often ineffective, costly, and risky. Existing systems not initially designed with safety considerations frequently require extensive redesigns to adequately incorporate essential safety mechanisms. Safety-by-design principles, such as those in ISO 26262 for automotive systems or IEC 61508 for industrial applications, are intended to guide development from inception. For automotive work, also check ISO 16750 (environmental), ISO 6469 (EV safety), ISO 11452 (EMC), and ISO 12405 (battery tests), and cite the specific part you applied; standards evolve, so always confirm the current edition before reuse.

Retrofitting often reveals uncomfortable truths: undocumented assumptions, inadequate requirements traceability, and architectural decisions that fundamentally conflict with safety principles. The deeper into development a project progresses before applying safety processes, the more expensive and disruptive remediation becomes—often increasing by an order of magnitude in cost when detected in later phases.

Example 8.5: *Consider a medical infusion pump initially developed as a standard commercial product that a company now wants to market for high-risk applications requiring safety certification.*

Myth-based Approach: *The company assumes they can simply document the existing design, run additional tests, and obtain safety certification without significant changes.*

Reality-based Challenges:

- ***Missing Design Artifacts:*** *The original hazard analysis and risk assessment were never performed, making it impossible to verify whether all risks were considered*
- ***Architectural Issues:*** *The single-processor design lacks independent monitoring for critical functions, requiring fundamental hardware changes*
- ***Software Development:*** *Source code was not developed according to safety standards, with inadequate requirements traceability and testing coverage*
- ***Undefined Requirements:*** *Critical safety requirements were never explicitly defined or validated*
- ***Cost Implications:*** *Retrofitting safety processes costs four to eight times more than integrating them from the beginning*

The company ultimately faces three undesirable options: (1) perform a costly complete redesign, (2) severely limit the product's intended use to avoid safety requirements, or (3) attempt partial remediation while accepting significant residual risk.

This example illustrates why safety cannot be effectively "bolted on" after development and requires integration from the earliest design stages.

Why It's Uncomfortable: Retrofitting safety processes obscures original design rationales, introduces significant project risks, and typically incurs higher overall costs than proactive integration. It places organizations in difficult positions of either accepting significant technical debt or undertaking expensive redesigns, neither of which is appealing to management or customers.
Takeaway: Embed safety considerations early and consistently throughout the design lifecycle. If retrofitting is unavoidable, transparently document assumptions, constraints, and associated risks. Continuously quantify and monitor the risks involved in backfilling processes. Consider approaches like safety wrappers, architectural boundaries, and operational limitations to mitigate risks when full redesign isn't feasible.

8.5 Safety Teams

The Myth: Safety teams inherently align with engineering and management goals.
The Reality: Safety teams often find themselves in adversarial relationships with engineering groups due to divergent priorities—engineers emphasize functionality, whereas safety teams focus on mitigating risk. Conflicts can arise when safety requirements impact schedules or budgets. This tension manifests in various ways: dismissal of safety concerns as theoretical, pressure to approve designs despite unresolved issues, or marginalization of safety personnel in decision-making processes.

Organizational structures exacerbate this problem when safety teams report to the same management chain responsible for delivery schedules, creating inherent conflicts of interest. Independence is crucial but difficult to maintain, especially when project timelines or budgets are threatened by safety considerations.
Why It's Uncomfortable: This tension can lead to organizational silos, with safety perceived as obstructive rather than beneficial. Safety teams may become isolated or pressured into compromising their standards. In extreme cases, safety professionals face ethical dilemmas between maintaining professional integrity and supporting organizational goals.
Takeaway: Cultivate collaboration by aligning safety objectives with overarching project goals. Maintain open, clear communication to bridge divides and articulate the value of safety efforts. Mature leadership in safety management is needed to navigate conflicts effectively. Implement organizational structures that preserve safety team independence while facilitating constructive engagement with development teams. Create incentives that reward proactive safety integration rather than treating safety as a checkbox exercise.

8.6 Customers

The Myth: Customers always prioritize safety above other concerns.
The Reality: Customers often prioritize cost, performance, and timelines over stringent safety measures, especially if they are unfamiliar with the complexities of safety-critical systems. They may underestimate the infrastructure or training necessary for proper system operation and maintenance. This priority mismatch becomes particularly evident in competitive bidding situations where adding comprehensive safety features may price a solution out of consideration.

Customer expectations sometimes conflict with safety best practices, such as demanding rapid feature additions that bypass proper verification processes or requesting simplified interfaces that remove important safety interlocks. Healthcare, industrial automation, and defense sectors all demonstrate instances where purchasing decisions favored cost or capability over safety considerations.
Why It's Uncomfortable: Safety professionals face pressure to minimize safety requirements or reduce training to meet customer expectations, potentially compromising overall system safety. This creates ethical dilemmas when business imperatives clash with professional safety obligations. Competitors willing to make safety compromises may gain market advantage, creating industry-wide pressure to reduce safety margins.
Takeaway: Educate customers on the long-term benefits of investing in robust safety measures. Transparently balance safety requirements with cost, performance, and delivery considerations, highlighting the risks of inadequate safety infrastructure and training. Develop strategies to clearly articulate safety value propositions about reduced lifecycle costs, operational reliability, and protection from liability. Industry standards and regulations can help level the playing field by establishing minimum safety requirements all competitors must meet.

8.7 Most Work Equals Least Results

The Myth: Every safety-related activity improves overall system safety. Figure 8.3 sketches diminishing returns: activity without signal does not translate into safety outcomes.
The Reality: Many resource-intensive activities, such as excessive documentation or detailed analyses of low-risk scenarios, have minimal actual impact on safety. Overemphasis on low-impact tasks can divert attention and resources away from genuinely critical issues. The Pareto principle often applies—approximately 20% of safety activities address 80% of the actual risk.

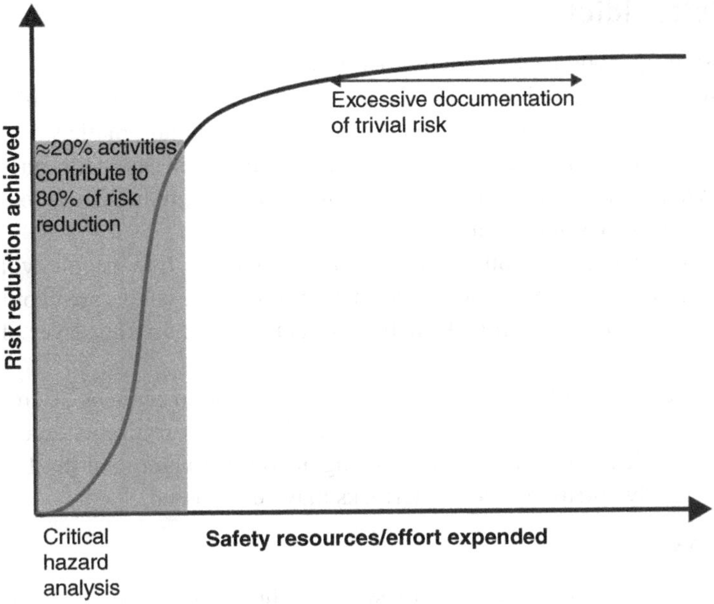

Figure 8.3 The safety activity efficiency curve.

Compliance-driven activities may consume disproportionate resources while yielding limited safety benefits. For example, exhaustive FMEA analyses of non-critical components, voluminous hazard logs with trivial entries, and comprehensive traceability matrices for low-risk requirements can create a facade of thoroughness without meaningful risk reduction.

Why It's Uncomfortable: Organizations often reward visible effort rather than meaningful contributions to safety. Safety processes become ritualized without critical evaluation of their effectiveness. Questioning established practices risks being perceived as undermining safety, even when those practices have dubious value.

Takeaway: Prioritize high-impact safety activities. Regularly evaluate and adjust safety processes to ensure resources focus effectively on critical risks, avoiding unnecessary complexity and effort. Implement risk-based approaches that concentrate verification and validation efforts on the most safety-critical aspects of systems. Consider safety metrics that measure actual risk reduction rather than process compliance alone. Encourage a culture that values effectiveness over volume in safety activities.

8.8 The Better Idiot

The Myth: Designing "foolproof" systems ensures absolute safety.
The Reality: Systems perceived as foolproof can encourage complacency or misuse, prompting users to engage in riskier behavior. Reliance on automatic safety features might inadvertently reduce vigilance and manual checks. This phenomenon, known as risk compensation or the Peltzman effect, manifests across domains from automotive safety to medical devices.

When you see this pattern in your own work, pause and ask how people will really behave around the system. A brief, honest paragraph in the safety case about expected human shortcuts often unlocks better mitigation than another layer of automation alone.

Ironically, adding safeguards sometimes introduces new failure modes or creates operational complexities that increase error likelihood. For example, excessive warning systems lead to alarm fatigue, causing operators to ignore or disable critical alerts. Similarly, intrusive safety interlocks may be bypassed.

Key Takeaways

- Recognizing and openly addressing these "dirty little secrets" matters for developing genuinely robust and resilient safety-critical systems.
- The most effective safety professionals acknowledge these realities rather than perpetuating myths of perfect systems.
- They build safety cultures that balance rigorous processes with practical effectiveness, being transparent about limitations while striving for excellence.

Part V

Management

9

Managing Projects

After reading this chapter, you should be able to:

- Write one measurable, testable safety requirement.
- Link a hazard to a requirement and to a test in a small slice.
- Mark vague words and replace them with numbers or conditions.

Managing safety-critical projects requires technical understanding, organizational skills, and careful decision-making to uphold safety standards while achieving project goals. Unlike traditional projects that prioritize budgets or deadlines, safety-critical projects emphasize adherence to standards, thorough documentation, and structured processes. This chapter covers practical approaches for managing these projects.

9.1 Introduction to Safety-critical Project Management

Safety-critical project management is distinguished by its emphasis on risk mitigation and safety assurance throughout the project lifecycle. The fundamental challenge lies in balancing three competing constraints: safety requirements, schedule pressures, and resource limitations—while ensuring safety always remains the nonnegotiable priority.

Good management requires recognizing that safety-critical projects operate under regulatory frameworks and certification requirements that influence project structure, documentation standards, and verification approaches.

9.2 Risk Management

For V&V rhythms that surface risk early, see Section 5.4.

Risk management in safety-critical projects systematically addresses risks that could have severe safety implications. See Figure 9.1 for a compact view of balancing safety, cost, and schedule pressure.

9.2.1 Key Principles

Risk management in safety-critical projects rests on three principles that, when applied consistently, improve the likelihood of delivering safe systems on schedule.

Early risk identification forms the cornerstone of the approach. The cost and difficulty of addressing risks increases dramatically as projects progress, so identifying hazards early creates the maximum opportunity for effective mitigation. Teams should employ techniques like preliminary hazard analysis (PHA), fault tree analysis (FTA), and failure modes and effects analysis (FMEA) early in the project lifecycle to uncover potential risks. This identification effort should cover both technical risks (hardware failures, software defects) and process-oriented risks (verification delays, resource shortages, supplier issues) that could affect the project's ability to deliver safely.

Clear risk prioritization ensures that limited resources focus on the most significant threats. A clearly defined risk matrix that categorizes risks by both likelihood and severity enables teams to make rational resource allocation decisions. High-severity risks deserve attention even when their perceived likelihood is low—the potential consequences justify preventive investment. Without explicit prioritization, teams often gravitate toward addressing familiar or easily resolved risks while more serious threats remain neglected.

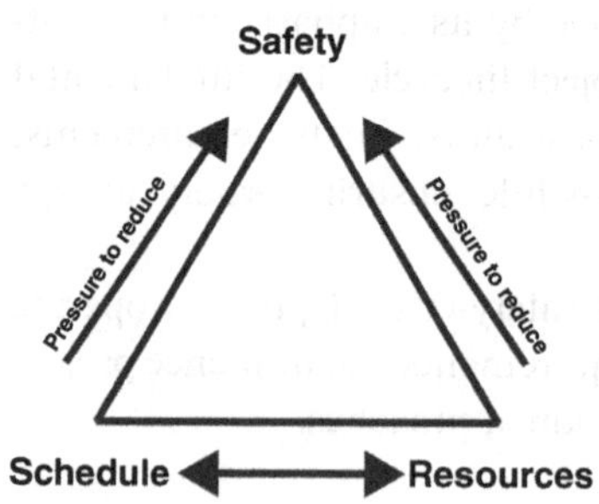

Figure 9.1 Safety-critical project management triangle.

Continuous monitoring and updates acknowledges that risk landscapes evolve as projects progress. New information emerges from testing, design decisions create or eliminate hazards, and external circumstances change. Teams should regularly revisit risk assessments throughout the project to capture new risks and update mitigation strategies as understanding deepens. Test results, integration findings, and operational insights should all feed back into the risk management process, keeping it current and relevant.

9.2.2 Risk Management Example

Consider a medical infusion pump development project, where a key risk involves the software control algorithm that regulates medication flow rates:

```
Risk: Control algorithm fails to regulate medication flow
rate within safe limits
Severity: Critical (could cause patient harm through
incorrect medication dosing)
Likelihood: Medium (based on complexity of algorithm and
novel design elements)

Mitigation Strategies:
1. Implement independent monitoring system for flow
   rate verification
2. Add hardware flow limiters as secondary protection
   mechanism
3. Develop comprehensive verification protocol specifically
   for flow rate accuracy
4. Design fault detection algorithm with appropriate alarms
   and safe shutdown mode
```

This example demonstrates proper risk analysis by identifying a specific risk, evaluating its severity and likelihood, and developing multiple layers of protection to ensure patient safety is maintained even if a single system fails.

9.2.3 Practical Advice

- Balance rigorous risk analysis with project momentum to avoid analysis paralysis.
- Maintain detailed documentation of risk decisions and rationale for auditability and regulatory compliance.
- Establish clear risk escalation pathways to ensure appropriate visibility of safety-critical issues.

9.3 Schedule Management

Schedule management in safety-critical projects demands careful coordination due to complex interdependencies, especially within verification and validation (V&V).

9.3.1 Key Concepts

Three scheduling concepts prove particularly important in safety-critical project management, each addressing a different aspect of the complexity inherent in these projects.

Task dependencies require explicit identification and management in safety-critical work. Unlike conventional projects where many activities can proceed in parallel, safety-critical projects often have strict sequencing requirements—verification cannot begin until design is complete, integration cannot proceed until components are verified, and regulatory submissions depend on completed evidence packages. Managers must distinguish clearly between tasks that can occur in parallel and those that are strictly sequential, then explicitly manage these dependencies to minimize unexpected delays. A missed dependency can cascade through the schedule, creating delays far larger than the original slip.

Critical path method (CPM) analysis helps identify which tasks directly drive project duration. Critical path tasks are those whose delays will directly impact overall project timelines—they have no schedule slack. In safety-critical projects, the critical path often runs through verification and validation activities or regulatory approval processes rather than through development work. Managers should ensure adequate resource allocation and oversight for critical path activities, recognizing that these tasks deserve priority attention and protection from competing demands.

Contingency buffers provide essential protection against the uncertainties inherent in safety-critical work. Testing may reveal defects requiring redesign. Regulators may request additional evidence. Analysis may uncover hazards requiring mitigation. Realistic time buffers should be built into the schedule to account for these unforeseen delays, especially during critical testing and regulatory approval phases where rework is most common and most expensive. Schedules that assume everything will go perfectly the first time are schedules that will inevitably slip.

9.3.2 Safety-critical Milestones

Safety-critical projects require specific milestones beyond those in conventional projects:

- Safety requirements review
- Preliminary hazard analysis completion
- Design safety review
- Safety case preparation
- Independent safety assessment
- Regulatory submission
- Certification/approval

These milestones should be explicitly integrated into the project schedule with appropriate review gates and dependencies.

9.3.3 Practical Advice

- Integrate safety-critical tasks explicitly into the project schedule for visibility and prioritization.
- Regularly update schedules during milestone reviews to accurately reflect current progress and forecast challenges.
- Use schedule visualization techniques that highlight safety-critical paths and dependencies.

9.4 Temptations to Avoid

Example 9.1: ***Schedule-driven Risk Masking.*** *A team green-lights integration to hit a milestone while three safety-critical assumptions remain unverified. The dashboard shows "on track," but residual risk is invisible. Fix: require an explicit "assumption ledger" with owners and due dates, and gate milestones on closure or accepted residual risk with sign-off.*

High-pressure situations can tempt project managers into shortcuts that compromise safety.

9.4.1 Common Temptations

High-pressure situations create temptations to take shortcuts that may seem reasonable in the moment but ultimately compromise safety. Recognizing these temptations helps managers resist them.

Delaying safety analyses often seems attractive when schedules are tight. The reasoning—that safety concerns can be resolved later in the development cycle—appears to preserve progress while deferring difficult work. In reality, late-stage analyses frequently reveal significant issues that require costly design changes, sometimes invalidating months of development effort. The later hazards are discovered, the more constrained the options for addressing them, and the greater the risk of exposure until they are resolved.

Over-documentation represents a different kind of trap. The misconception that extensive documentation always enhances safety can lead to massive documentation efforts that consume resources without improving actual safety. When documentation becomes excessive, critical information gets buried in volumes of text that nobody reads carefully. The effort spent maintaining documentation could have been spent on analysis, testing, or design improvement. Effective documentation is concise and purposeful, not comprehensive for its own sake.

Reliance solely on testing reflects a misunderstanding of what testing can achieve. The belief that comprehensive testing alone guarantees system safety ignores a fundamental limitation: testing can reveal the presence of defects but cannot prove their absence. Testing identifies existing issues in the scenarios actually tested, but it cannot eliminate the possibility of latent or emergent faults that activate under conditions the tests never explored. Safety assurance requires complementary methods beyond testing.

Compromising safety for deadlines is perhaps the most dangerous temptation. When schedules slip, pressure mounts to declare work complete or to skip verification steps that seem unlikely to find problems. The reasoning that meeting deadlines takes precedence over thorough safety evaluations treats schedule as paramount and safety as negotiable. In reality, cutting corners in safety evaluations substantially heightens the risk of severe operational failures—failures whose costs in lives, liability, and reputation far exceed any schedule benefit.

9.4.2 Practical Advice

- Clearly define nonnegotiable safety milestones and uphold them rigorously.
- Regularly reassess and challenge project assumptions to avoid inadvertently introducing unacceptable risks.
- Foster a culture where raising safety concerns is rewarded rather than penalized.

9.5 Viable Efficiency Strategies

Efficiency strategies can optimize project progress without compromising safety standards. Figure 9.2 shows how model-based development front-loads verification and improves traceability.

9.5.1 Effective Strategies

Several strategies can improve efficiency in safety-critical projects without compromising safety standards. Each addresses a different source of inefficiency while maintaining the rigor that safety demands.

Modular testing enables parallel development paths that would otherwise be impossible. By designing systems with well-defined interfaces and independently testable subsystems, teams can verify components in parallel rather than waiting for full system integration. This approach reduces integration complexity and time while providing earlier feedback on component quality. When subsystems are verified before integration, integration testing can focus on interface issues rather than component defects.

Utilization of proven components leverages previous verification investments. Reusing well-established designs and certified components accelerates development by avoiding the need to verify functionality that has already been demonstrated in other applications. The key is ensuring that the proven component is being used within its validated operating envelope—a component verified for one application may not be suitable for another without additional analysis. When applicable, this strategy improves both reliability and schedule.

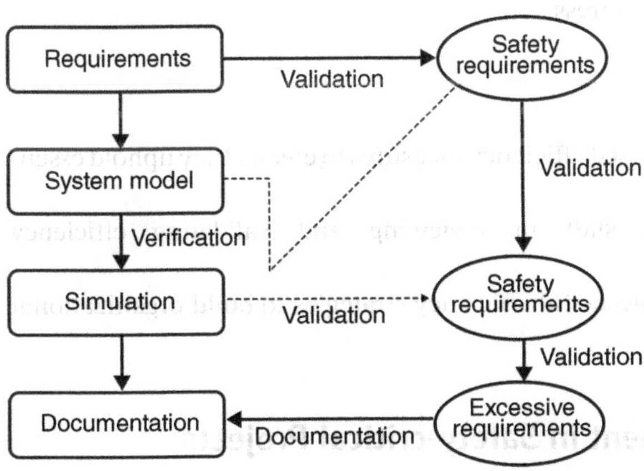

Figure 9.2 Model-based development in safety-critical projects.

Automated testing and analysis increases verification throughput without sacrificing rigor. Automated tools for software verification, static analysis, and documentation generation can execute far more tests, check far more code properties, and generate far more consistent documentation than manual processes. The efficiency gains free human experts to focus on judgment-intensive activities like hazard analysis and design review, while routine verification runs automatically. The key is selecting and configuring tools appropriately—automation amplifies whatever is encoded, whether good practice or flawed assumptions.

Incremental verification and validation provides earlier feedback and reduces rework. Rather than waiting until systems are complete to begin verification, teams can verify and validate subsystems and features incrementally as they become available. This approach identifies and resolves issues early, when fixes are less expensive and less disruptive. It also provides continuous visibility into system quality, enabling better-informed decisions about schedule and resource allocation.

9.5.2 Model-based Development

Model-based development can improve efficiency in safety-critical projects by:

- Enabling early validation through simulation before physical implementation
- Supporting automatic code generation from verified models
- Facilitating consistent documentation derived directly from models
- Improving traceability between requirements, design, and implementation

This approach front-loads verification activities, allowing issues to be identified when they are less costly to address.

9.5.3 Practical Advice

- Critically evaluate proposed efficiency measures to ensure they uphold essential safety requirements.
- Leverage experienced staff in reviewing and validating efficiency enhancements.
- Document and share successful efficiency strategies to build organizational knowledge.

9.6 Team Management in Safety-critical Projects

Managing diverse specialist teams well matters for safety-critical project success.

9.6.1 Cross-functional Collaboration

Safety-critical projects involve multiple disciplines, including:

- System engineers
- Safety engineers
- Software developers
- Hardware engineers
- Quality assurance specialists
- Regulatory affairs experts

Each discipline brings essential expertise but may also introduce communication challenges and conflicting priorities. Project managers must establish clear interfaces between teams and ensure safety information flows effectively across organizational boundaries.

9.6.2 Independence and Oversight

Safety assessment functions require appropriate independence from development activities to ensure objective evaluation. However, this independence must be balanced with integration into the development process to ensure safety concerns are addressed effectively.

Project managers should:

- Establish clear reporting structures that maintain assessment independence.
- Ensure safety representatives participate in key development decisions.
- Create formal review processes where safety perspectives are explicitly considered.
- Resolve conflicts between development and safety priorities at appropriate management levels.

9.6.3 Practical Advice

- Provide cross-disciplinary training to help team members understand safety implications across domains.
- Implement regular safety-focused reviews with participation from all relevant disciplines.
- Create shared safety objectives that align all teams toward common goals.

9.7 Regulatory Compliance Management

Safety-critical projects typically operate under regulatory frameworks that impose specific requirements and processes.

9.7.1 Key Considerations

Successful regulatory compliance requires proactive engagement and systematic planning, not reactive documentation assembly at project end.

Early regulatory engagement establishes shared understanding before significant design commitments are made. Engaging with regulators early helps teams understand expectations and requirements while there is still flexibility to address them efficiently. Many regulatory bodies welcome pre-submission meetings or informal consultations that can identify potential compliance challenges before they become expensive problems. Planning for regulatory submissions as integral project milestones ensures that compliance activities receive appropriate schedule and resource allocation rather than being treated as afterthoughts.

Compliance strategy provides the roadmap for demonstrating that regulatory requirements have been met. A clear strategy identifies which standards apply, how compliance will be demonstrated for each requirement, and what evidence will be generated. Tracing regulatory requirements to specific project deliverables and verification activities creates the linkages that reviewers need to assess compliance efficiently. Without this traceability, regulatory submissions become exercises in hunting for evidence rather than presenting organized arguments.

Documentation planning ensures that evidence generation is built into development processes rather than reconstructed after the fact. The documentation architecture should be designed to support regulatory submissions from the outset, with clear definitions of what documents will be produced, how they relate to each other, and how they demonstrate compliance. When evidence generation is integrated into routine development activities, it becomes an automatic byproduct of doing the work rather than a separate burden.

9.7.2 Practical Advice

- Assign explicit responsibility for regulatory compliance management
- Maintain awareness of evolving regulations and standards that may affect the project
- Build relationships with regulatory authorities based on transparency and credibility

Key Takeaway

- Cultivating a culture of transparency, accountability, and continuous improvement helps ensure that safety remains the top priority throughout the project lifecycle.

10

Managing People

After reading this chapter, you should be able to:

- Sketch a simple block view with the safety-relevant parts called out.
- State one interface rule and one isolation rule that matter for safety.
- Place a few safety requirements on blocks or interfaces and say why.

In safety-critical projects, managing personnel well is just as important as managing technical aspects. A common challenge is the inherent adversarial dynamic that often arises between engineering teams and the verification and validation (V&V) or safety groups. This chapter covers strategies for managing these dynamics, building collaboration, and ensuring that the organization remains collectively focused on delivering safe and reliable systems. People can also be a direct safety risk (insider threat, mistakes under pressure), so personnel practices are part of the safety case, not an afterthought.

Think of this chapter as the people layer of your safety argument. The lists below are checklists, not verdicts. Add short notes on what already works well in your organization and where you expect friction; that humility makes the guidance more usable.

10.1 Understanding the Adversarial Dynamic

For roles and independence boundaries, see Section 2.1.

The Engineering Perspective:

- Engineers typically prioritize functionality, performance, innovation, and deadlines.
- Safety and V&V requirements are often perceived as hindrances that delay schedules or complicate designs.

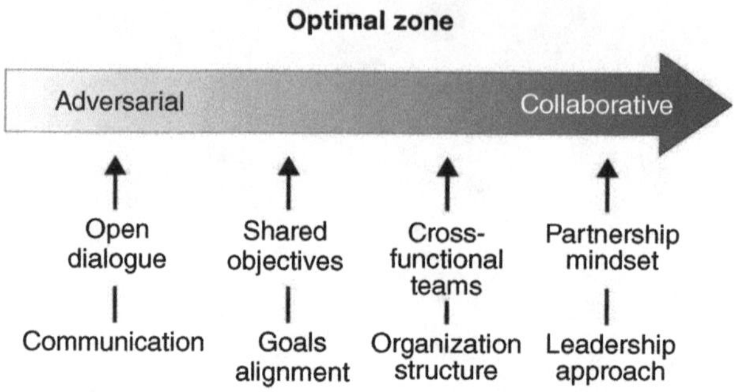

Figure 10.1 The engineering-safety relationship spectrum.

See Figure 10.1 for a simplified spectrum from adversarial to collaborative postures.

The Safety and V&V Perspective:

- V&V teams focus on identifying, evaluating, and mitigating risks, ensuring compliance with stringent safety standards.
- Engineering decisions may appear rushed or inadequately justified, prompting increased oversight and scrutiny from safety teams.

Reasons for This Dynamic:

- *Diverging Goals:* Engineering aims for rapid delivery of functional systems; safety teams emphasize thorough risk reduction and compliance.
- *Different Metrics of Success:* Engineers measure success by functionality and innovation, while safety teams measure success through reduced risk and verified compliance.
- *Communication Barriers:* Different terminologies, methodologies, and priorities lead to misunderstandings and communication gaps.
- *Organizational Structure:* Traditional hierarchies may inadvertently reinforce separation rather than integration of these functions.

10.2 Organizational Structure Considerations

The reporting structure affects how engineering and safety teams interact. Several models exist, each with distinct advantages and challenges.

10.2.1 Independent Safety Authority

In this model, safety teams report to a separate organizational authority from engineering teams, maintaining independence for objective assessment. This structure reflects the principle that those responsible for evaluating safety should not be subject to the same pressures as those responsible for delivering the product.

The primary advantage of this structure is that stronger independence allows unbiased safety assessments. When safety personnel do not report to managers whose performance is measured by delivery schedules, they face less pressure to compromise safety standards for schedule. The clear separation of concerns also reinforces safety's organizational importance, signaling that the organization takes safety seriously enough to invest in dedicated oversight.

However, organizational separation carries real costs. The structure may reinforce an "us versus them" mentality, with safety perceived as adversarial rather than collaborative. Decision-making processes can become slower when approval must traverse organizational boundaries. Perhaps most importantly, separation creates potential for communication breakdowns—safety teams may lack the detailed design knowledge needed to assess risks effectively, while engineering teams may not understand the rationale behind safety requirements. Managing these challenges requires deliberate effort to maintain constructive relationships across the organizational divide.

10.2.2 Matrix Organization

In a matrix organization, safety personnel have dual reporting lines: functionally to a safety director who ensures technical standards and career development, and operationally to project managers who direct day-to-day work. This structure attempts to capture the benefits of both integration and independence.

The matrix approach offers several advantages. It balances independence with project integration, keeping safety personnel close enough to development work to understand it deeply while maintaining a separate functional home. Consistent safety practices can be enforced across projects because the safety director oversees all safety personnel regardless of their project assignments. The dual reporting relationship fosters both technical excellence (through the functional line) and project awareness (through the operational line).

The challenges of matrix organization center on the complexity of dual reporting. Safety personnel may receive conflicting priorities and direction from their two managers—the safety director emphasizing thoroughness while the project manager emphasizes schedule. Strong coordination mechanisms become essential to resolve these conflicts without putting safety personnel in impossible positions. The structure can also create ambiguity in decision authority, with neither manager clearly empowered to make final calls. Successful matrix organizations invest heavily in defining how conflicts will be resolved and ensuring both reporting lines respect each other's legitimate concerns.

10.2.3 Practical Implementation

Regardless of the chosen structure, effective organizations typically incorporate:

- Clear escalation paths for unresolved safety concerns
- Regular joint reviews between engineering and safety leadership
- Documented independence requirements for critical safety decisions

10.3 Strategies for Managing the Dynamic

Case in Point. A program introduced a weekly "risk sync" (30 minutes, single owner, shared board) where engineering and safety each bring one *new* concern and one *closed* item. Within two sprints, duplicate work dropped, evidence requests became specific, and a previously contentious review closed without escalation. The change was small, observable, and reinforced by leadership attending the first two sessions.

10.3.1 Establish Shared Goals

Clearly articulate that both teams ultimately share the same primary objective: a safe and reliable system. Emphasize safety as an enabler of robust, reliable system functionality rather than an impediment.

Action Steps:

- Develop a unified project vision document incorporating both safety and engineering priorities.
- Regularly highlight joint achievements in team meetings to foster mutual respect and shared ownership.
- Create combined performance metrics that balance development progress with safety adherence.

10.3.2 Foster Open Communication

Good communication reduces misunderstandings and builds mutual trust between teams.

Action Steps:

- Schedule regular, structured cross-team meetings to discuss progress, risks, and challenges openly.
- Establish shared documentation platforms to ensure unified access to critical project information.

- Create a common glossary of terms to align terminologies used by engineering and safety teams.
- Implement "early warning" mechanisms where teams can flag potential issues before they become formal problems.

Communication Example: Consider a situation where an engineering team needs to modify a critical algorithm controlling a medical device's drug delivery. A communication protocol might include:

```
1. Engineering provides initial notification of intended
   change
2. Brief discussion at weekly cross-team meeting to outline:
 - Reason for the change
 - Potential safety implications
 - Verification approach
3. Safety team provides feedback on risk assessment needs
4. Joint documentation of change impact and verification
   strategy
5. Engineering implements with agreed safety checkpoints
6. Safety team verifies against agreed criteria
```

This structured approach ensures transparency, appropriate involvement, and shared responsibility without unnecessary delays or conflicts.

10.3.3 Define Clear Roles and Responsibilities

Clearly defined roles prevent overlap and misunderstandings about who is responsible for specific tasks, improving accountability and efficiency.

Action Steps:

- Use tools such as RACI matrices (responsible, accountable, consulted, informed) to clarify each team's responsibilities.
- Clearly document responsibilities:
  - Engineering Team: Responsible for functional designs and performance.
  - Safety Team: Responsible for compliance, hazard identification, and risk management.
  - Both Teams: Consulted on any significant changes affecting project requirements or processes.
- Review role definitions periodically to address gaps or overlaps.

10.3.4 Build Interdisciplinary Teams

Integrating personnel across functional boundaries builds empathy, mutual understanding, and improved teamwork.

Action Steps:

- Assign safety representatives as liaisons within engineering teams to provide ongoing safety guidance.
- Offer opportunities for personnel exchanges, rotating engineers through V&V roles and vice versa.
- Conduct joint training sessions where both teams learn together.
- Create interdisciplinary task forces for specific challenges requiring multiple perspectives.

10.3.5 Manage Conflicts Constructively

Conflicts are inevitable, but constructive management ensures issues are addressed productively without damaging relationships.

Action Steps:

- Establish clear, structured conflict resolution processes:
  - Clearly identify and define issues and their implications.
  - Base discussions and decisions on factual data rather than opinions or assumptions.
  - Identify common goals to facilitate collaborative solutions.
  - Involve neutral mediators, such as project managers, when conflicts persist.
- Document decisions and rationales to prevent recurring arguments on settled issues.
- Build reflection opportunities to learn from past conflicts.

10.3.6 Provide Training and Support

Training ensures teams have the necessary knowledge and skills to collaborate effectively and understand each other's priorities and methodologies.

Action Steps:

- Conduct targeted safety training for engineers, highlighting the importance and practical integration of safety standards.
- Provide V&V teams with collaboration training, focusing on constructive feedback and solution-oriented discussions.
- Develop specialized courses on the interface between engineering and safety disciplines.
- Offer mentoring programs where experienced staff guide newer personnel in cross-team collaboration.

10.3.7 Recognize and Reward Collaboration

Positive reinforcement can alter team dynamics and encourage ongoing collaboration.

Action Steps:

- Implement recognition programs explicitly designed to reward cross-team collaboration.
- Publicly acknowledge successful cooperative efforts to reinforce positive behaviors and collaborative spirit.
- Include collaboration effectiveness in performance reviews and promotion criteria.
- Share success stories of effective collaboration throughout the organization.

10.4 Human Factors in Team Dynamics

Understanding the psychological aspects of team interaction can help address root causes of adversarial dynamics.

10.4.1 Cognitive Biases

Several cognitive biases can influence how engineering and safety teams perceive each other, often in ways that exacerbate rather than resolve conflicts. Understanding these biases helps managers recognize when they are operating and intervene constructively.

Confirmation bias leads teams to emphasize information that confirms their existing perspectives while discounting contradictory evidence. Engineers who believe safety teams are obstructionist will notice every delay caused by safety review while overlooking instances where safety input prevented costly errors. Safety teams who believe engineers cut corners will focus on every deviation from standards while ignoring the many cases where engineering exercised appropriate judgment.

In-group favoritism is a natural human tendency to favor members of one's own group and view other groups skeptically. Engineering teams develop shared identity around solving technical problems; safety teams develop shared identity around protecting against harm. Each group may come to see the other as outsiders whose concerns are less legitimate than their own. This bias operates automatically and unconsciously, making it particularly difficult to counteract.

Fundamental attribution error involves attributing others' actions to their character rather than situational factors. When safety review delays a milestone, engineers may conclude that safety personnel are inherently obstructionist rather than recognizing they may be responding to genuine concerns or resource constraints. When engineers push back on requirements, safety personnel may conclude they don't care about safety rather than recognizing they may have legitimate technical objections or schedule pressures. This bias poisons relationships by transforming situational conflicts into perceived character flaws.

10.4.2 Addressing Human Factors

- Promote awareness of common biases through training and reflection.
- Create psychologically safe environments where team members can express concerns without fear.
- Implement structured decision-making processes that require examination of multiple perspectives.
- Rotate meeting leadership and facilitation roles between engineering and safety teams.

10.5 Leadership Approaches

Good leadership matters for fostering positive team dynamics in safety-critical environments.

10.5.1 Key Leadership Practices

- **Model Collaborative Behavior:** Leaders should demonstrate respect for both engineering and safety perspectives.
- **Maintain Consistent Safety Messaging:** Regularly reinforce that safety is nonnegotiable, regardless of schedule pressure.
- **Make Value-based Decisions:** Explicitly reference organizational values when resolving conflicts.
- **Balance Advocacy with Inquiry:** Leaders should both advocate for their perspective and genuinely inquire about others' viewpoints.

10.5.2 Practical Leadership Tools

- Regular skip-level meetings with both engineering and safety personnel
- Joint problem-solving sessions led by senior management
- Executive reviews that examine both technical progress and safety adherence
- Leadership development focused on managing technical-safety dynamics

10.6 Case Example: Automated Medical Device

The Challenge: A medical device development team designed a new patient monitoring system focusing on functionality and user-friendliness. The safety team identified risks associated with sensor accuracy under certain environmental conditions. Engineers initially perceived these safety concerns as exaggerated, leading to tension.

The Resolution:

- *Joint Risk Review Workshops:* Both teams collaboratively evaluated sensor reliability and environmental risks.
- *Collaborative Solutions:* Engineers introduced redundancy with additional sensors and improved calibration procedures, and the safety team validated these improvements.
- *Recognition:* Joint recognition was provided to both teams during organizational meetings, positively reinforcing collaborative behavior.

10.7 Measuring Team Effectiveness

To improve team dynamics, organizations need concrete ways to measure collaboration effectiveness.

10.7.1 Key Performance Indicators

- **Process Indicators:**
  - Frequency of joint engineering-safety meetings
  - Time to resolve cross-team issues
  - Number of collaboratively developed solutions
- **Outcome Indicators:**
  - Reduction in late-stage safety issues
  - Decreased rework due to safety considerations
  - Improved system safety metrics with maintained development efficiency
- **Cultural Indicators:**
  - Team member satisfaction surveys
  - Self-reported psychological safety scores
  - Willingness to raise cross-functional concerns

10.7.2 Regular Assessment Process

Implement a structured process to periodically evaluate team dynamics:

- Quarterly reviews of collaboration metrics
- Regular retrospectives involving both engineering and safety teams
- Annual comprehensive assessment with improvement planning

Key Takeaways

- Building a culture of mutual respect, accountability, and collaboration improves both project efficiency and system safety.
- The most successful safety-critical organizations recognize that technical excellence and safety assurance are complementary, not competing, objectives.
- They invest in organizational structures, leadership development, and team processes that bridge traditional divides between engineering and safety functions.

Part VI

Deployment and Support

11

Supporting New Systems

After reading this chapter, you should be able to:

- Write a short incident flow: contain, capture evidence, communicate.
- Do a quick root-cause and record one action that sticks.
- Add one leading indicator that shows drift before it bites.

Once a safety-critical system becomes operational, ongoing monitoring and maintenance matter for preserving safety and reliability. Good monitoring allows for early identification of issues, systematic data collection, prompt corrective actions, and continuous system improvements. This chapter covers guidance and templates for tracking and maintaining systems post-deployment, helping organizations predict issues, resolve problems, and sustain safety. Think of this as the operational half of the safety case: vigilance, evidence capture, and timely correction.

Templates and lists are only useful if the team believes they save time. Tailor each template, keep examples close at hand, and write down one or two "do not skip" items for your context. It will feel more like a teammate's advice and less like bureaucratic overhead.

11.1 Why Monitor?

Deployment marks the beginning, not the end, of safety responsibility. Operational monitoring serves multiple purposes that together ensure the safety case remains valid throughout the system's operational life. See Figure 11.1 for a compact feedback loop from observation to corrective action.

Ensuring continued safety is the most fundamental purpose of monitoring. The verification and validation completed before deployment demonstrated safety under test conditions, but real-world operation exposes the system to conditions

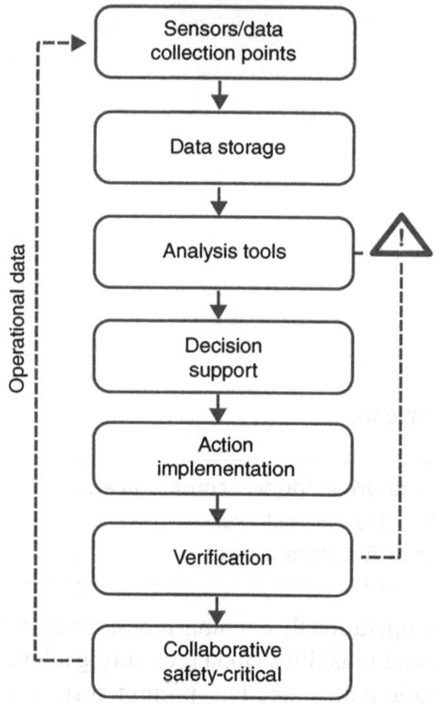

Figure 11.1 The monitoring feedback loop in safety-critical systems.

that testing could not fully replicate. Monitoring confirms that the system operates safely under actual operational conditions and enables quick detection and response when potential failures emerge. Without this ongoing verification, the safety case developed during development gradually becomes stale as the gap between tested conditions and actual conditions widens.

Identifying latent issues becomes possible only through operational observation. Some defects are not apparent during initial testing—they may require longer exposure times, specific environmental conditions, or particular usage patterns to manifest. Operational monitoring can reveal these hidden defects before they cause harm. It also captures system responses in unexpected or untested scenarios that inevitably arise in real-world use, providing valuable information about the system's actual behavior envelope.

Supporting compliance requires ongoing evidence generation. Regulatory frameworks increasingly require documented evidence of operational safety, not just pre-deployment verification. Monitoring provides this evidence for regulatory audits and demonstrates continuous adherence to applicable safety standards. Organizations that cannot demonstrate ongoing safety monitoring may face regulatory challenges even if their systems perform well.

Improving future designs leverages operational experience to make next-generation systems safer and more reliable. Monitoring gathers operational insights that inform enhancements in future system iterations, identifying persistent weaknesses in design or components that should be addressed. This feedback loop from operations to design is essential for organizational learning—without it, each project starts from scratch rather than building on accumulated experience.

Establishing operational baselines enables detection of gradual changes that would otherwise go unnoticed. By developing performance baselines early in operation, teams can detect gradual degradation over time—drift that occurs too slowly to trigger immediate alarms but eventually compromises safety margins. These baselines also create reference points for evaluating the impact of system modifications or updates, distinguishing expected changes from unexpected degradation.

11.2 What to Monitor?

Good monitoring requires careful selection of what to observe. The goal is not to capture everything—that would create data overload—but to capture indicators that reliably signal safety-relevant changes. Several monitoring aspects are common across most safety-critical systems, though the specific metrics will vary by domain. Figure 11.2 illustrates layered monitoring from component signals to system-level outcomes.

11.2.1 System Performance Metrics

Performance metrics track whether the system continues to meet its operational specifications. Key examples include response times, accuracy of critical sensors, and reliability of key operational elements. The purpose is to ensure adherence to performance specifications under varying conditions—specifications that were validated during development and form part of the safety case. Monitoring frequency should be continuous for the most critical parameters, with defined intervals acceptable for less critical metrics based on how quickly degradation could progress to an unsafe state.

Example: Medical Infusion Pump Performance Monitoring

```
Performance Metric: Flow Rate Accuracy
Target: ±2% of programmed rate
Monitoring Method: Automated flow sensors with hourly
logging Alert Threshold: >±3% deviation for more than three
consecutive readings
Response Protocol: Immediate calibration check, removal
from service if persistent deviation detected
```

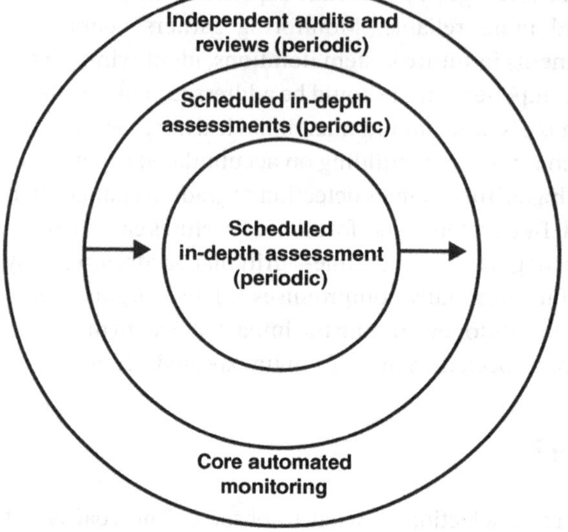

Figure 11.2 Layered monitoring approach for safety-critical systems.

11.2.2 Failure Modes and Faults

Monitoring failures and faults captures information about what goes wrong and when. Examples include instances of sensor malfunctions, software crashes, or communication disruptions—any event where the system does not behave as intended. The purpose extends beyond immediate incident response to identifying failure patterns and proactively addressing root causes. Analysis should track frequency, duration, and conditions surrounding failures, looking for correlations that reveal underlying issues. A single sensor failure may be random chance; a pattern of sensor failures under specific conditions indicates a systematic problem requiring systematic correction.

11.2.3 Environmental Conditions

Environmental monitoring captures conditions that may affect system performance. Temperature extremes, humidity levels, and weather events can all influence system operation in ways that may not have been fully characterized during development. The purpose is to link environmental variables with performance fluctuations or system failures, building understanding of the system's actual operating envelope. This information enables operational responses such as adjusting operational parameters or maintenance schedules based on environmental factors and feeds back into design understanding for future systems.

11.2.4 Maintenance and Repairs

Maintenance tracking captures information about every intervention performed on the system. Component replacements, calibration updates, and software patches should all be documented with enough detail to support later analysis. The purpose is to track maintenance frequency and costs to identify unreliable components or processes that may warrant design changes or supplier changes. This data also enables optimization of preventive maintenance schedules and spare parts inventory, reducing both maintenance costs and the risk of stockouts that could extend downtime.

11.2.5 Human Interactions

Monitoring human interactions reveals how people actually use the system, which may differ significantly from how designers expected. Operator interventions, user feedback, and incident reports all provide windows into the human side of system operation. The purpose is to identify user-experience issues and opportunities for training or interface improvements. Particular attention should be given to patterns of recurring operator errors or frequent override actions—these often indicate interface designs that do not match operator mental models or operational requirements that conflict with how people naturally work.

11.2.6 Near Misses and Precursors

Near misses and precursor events are perhaps the most valuable monitoring targets, precisely because they occur more frequently than actual incidents while revealing the same underlying weaknesses. These are situations where safety margins were reduced but no incident occurred—an alarm that was almost missed, a procedure that was almost skipped, a component that almost failed. The purpose is to identify systemic weaknesses before they result in actual incidents. Near misses provide opportunities for preventive action without the consequences of actual failures, making them invaluable leading indicators of safety risk.

11.3 How to Monitor?

Efficient monitoring combines automated tools and structured manual processes.

11.3.1 Automated Monitoring Systems

- **Sensors and Logs:**
  - Use built-in sensors and logging tools to capture continuous real-time data.
  - Implement data validation checks to identify sensor failures or inaccurate readings.

- **Event Triggers:**
  - Configure automated alerts for deviations from normal operational parameters.
  - Establish tiered alert levels based on severity and response urgency.
- **Secure Data Storage:**
  - Maintain secure long-term data storage for subsequent analyses and trend monitoring.
  - Implement appropriate data retention policies based on regulatory requirements.

11.3.2 Routine Inspections

- **Scheduled Audits:**
  - Perform regular inspections to validate automated data and capture issues sensors might miss.
  - Use trained personnel who understand system operation and potential failure modes.
- **Checklists:**
  - Use standardized inspection checklists to ensure thorough and consistent evaluations.
  - Update checklists based on emerging operational knowledge.

11.3.3 Predictive Maintenance

- **Data Analytics:**
  - Apply historical data analysis to predict component failures before they occur.
  - Look for patterns and trends that precede known failure modes.
- **Diagnostic Testing:**
  - Regularly conduct detailed checks on key system components during scheduled downtimes.
  - Use noninvasive testing where possible to minimize operational disruption.

11.3.4 Feedback Loops

- **User Reports:**
  - Establish straightforward channels for operators and users to report concerns or anomalies.
  - Create a culture that encourages reporting without fear of blame.

- **Incident Reviews:**
  - Systematically analyze and document issues to inform targeted improvements.
  - Focus on systemic factors rather than individual errors.

11.4 Data Analysis and Interpretation

Case in Point. A slow drift in reported wheel speed appears only in cold weather. Trend analysis flags a seasonal pattern; pattern recognition links it to a supplier batch of connectors. The corrective action replaces the connectors and adds a cold-soak diagnostic. Subsequent monitoring confirms the drift disappears, and the diagnostic does not raise nuisance alerts.

Collecting data is only valuable when paired with effective analysis and interpretation methods.

11.4.1 Trend Analysis

- Track performance metrics over time to identify gradual degradation or drift.
- Apply statistical process control techniques to distinguish normal variation from significant changes.
- Look for correlations between different parameters and system behaviors.

11.4.2 Pattern Recognition

- Identify recurring patterns that precede failures or near-misses.
- Use both automated algorithms and human expertise to recognize complex patterns.
- Establish normal operational profiles to highlight deviations more effectively.

Example: Pattern Recognition in Monitoring Data

```
Observed Pattern: Control system processing delays increase
briefly 3-4 hours before communication failures
Response: Implement automated monitoring of processing times
with alerts when delays exceed 20ms for more than 5 minutes
Preventive Action: Develop procedure to restart
communication subsystem when early warning pattern detected
Validation: Log confirms 85% reduction in communication
failures after implementation of early pattern detection
```

11.4.3 Risk-based Analysis

- Focus analytical resources on monitoring data related to highest-risk components.
- Prioritize investigation of anomalies associated with safety-critical functions.
- Implement more sophisticated analysis methods for critical systems.

11.5 Handling Corrections

For change-control steps after containment, see Section 3.1.

Good incident handling matters for maintaining safety and reliability: Figure 11.3 summarizes the incident workflow and decision points for safe resolution.

11.5.1 Incident Prioritization

- Use risk-based prioritization to quickly address critical issues.
- Consider both immediate safety impact and long-term implications.
- Develop clear criteria for differentiating severity levels.

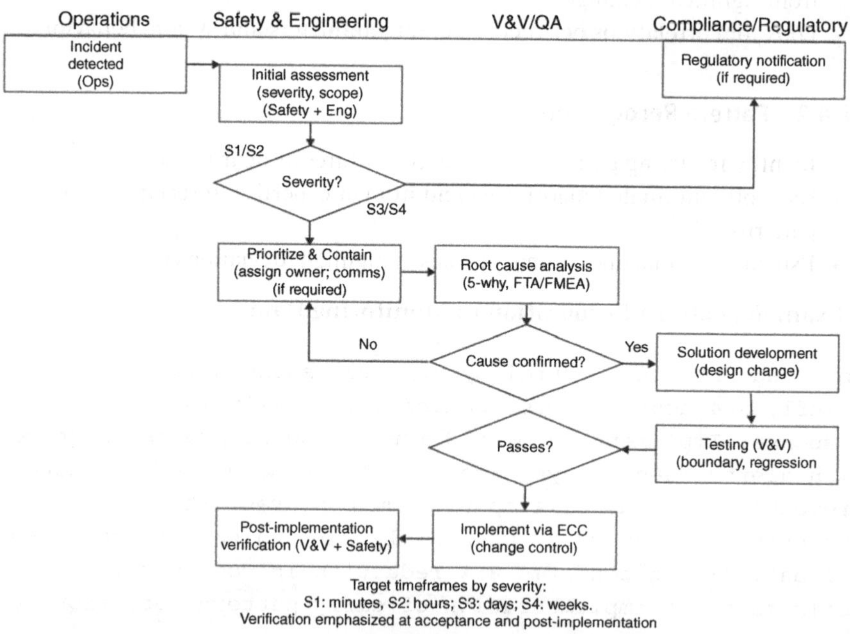

Figure 11.3 Incident response workflow for safety-critical systems.

11.5.2 Root Cause Analysis

- Apply structured RCA tools like fault tree analysis or fishbone diagrams.
- Document findings clearly for accountability and future reference.
- Look beyond immediate technical causes to examine organizational and systemic factors.

11.5.3 Implementation of Fixes

- **Short-term Mitigations:**
  - Quickly apply temporary solutions to mitigate immediate risks.
  - Document limitations of temporary measures and establish timelines for permanent solutions.
- **Long-term Solutions:**
  - Implement robust permanent corrections addressing root causes.
  - Consider potential side effects and unintended consequences of changes.

11.5.4 Validation and Verification

- Conduct controlled testing of fixes before broad deployment.
- Validate solutions thoroughly to confirm that no new hazards are introduced.
- Use appropriate verification techniques based on the nature and criticality of the change.

11.5.5 Documentation Updates

- Reflect all changes promptly in operational manuals and maintenance records.
- Ensure complete traceability of actions taken, their justifications, and validation outcomes.
- Update training materials to incorporate lessons learned and new procedures.

11.6 Establishing a Comprehensive Support System

A good support system for safety-critical systems requires several key components.

11.6.1 Support Team Structure

- Define clear roles and responsibilities for operational support personnel.
- Establish appropriate escalation paths for different severity levels.

- Ensure 24/7 coverage for critical systems with defined response times.
- Maintain expertise across all critical system components.

11.6.2 Knowledge Management

- Develop and maintain comprehensive system documentation.
- Create knowledge bases of known issues and solutions.
- Implement effective handover procedures between support shifts.
- Establish procedures for capturing and preserving institutional knowledge.

11.6.3 Continuous Improvement Process

- Regularly review incident patterns and response effectiveness.
- Conduct periodic support process audits and improvements.
- Update monitoring parameters based on operational experience.
- Incorporate feedback from operations and maintenance personnel.

11.7 Templates for Monitoring

Structured templates streamline the monitoring process.

11.7.1 System Performance Log

Date/Time	Metric	Recorded value	Expected range	Notes
2025-03-15	System response	3.4 seconds	2–5 seconds	Within range

11.7.2 Failure Incident Report

Incident ID	Date	Description	Root cause	Corrective action	Status
1001	2025-03-16	Sensor fault	Calibration drift	Recalibrated	Completed

11.7.3 Maintenance Record

Date	Component	Action taken	Reason	Next scheduled
2025-03-12	Actuator	Replaced	End-of-life	2026-03-12

11.7.4 Change Control Log

Change ID	Date	Description	Risk assessment	Verification method	Approval
CC-103	2025-04-02	Firmware update	Medium	Regression testing	J. Smith

11.7.5 Near-miss Report

Report ID	Date	Description	Contributing factors	Preventive action
NM-047	2025-03-22	Alarm delay	Network congestion	Prioritized traffic

11.8 Regulatory Considerations in Operational Support

Different industries have specific regulatory requirements for operational monitoring and support.

11.8.1 Industry-specific Requirements

- **Medical Devices:** FDA requirements for post-market surveillance and medical device reporting (MDR).
- **Aviation:** Continued airworthiness and service difficulty reporting (SDR) requirements.
- **Nuclear:** Maintenance rule (10 CFR 50.65) and reporting requirements for operational events.
- **Automotive:** Field monitoring and recall management under various regional regulations.

11.8.2 Compliance Documentation

- Maintain auditable records of operational performance and incidents.
- Document all maintenance activities according to regulatory standards.
- Prepare for regulatory inspections with organized evidence of compliance.
- Report significant incidents to appropriate authorities within required timeframes.

Key Takeaways

- Continuous monitoring and proactive maintenance matter for sustaining safety-critical systems post-deployment.
- Organizations that excel in operational support recognize that deployment marks the beginning, not the end, of their safety responsibility.

Part VII

Frontiers and Wrap-up

12

New Technologies

After reading this chapter, you should be able to:

- List allowed and not-allowed LLM uses in your project.
- Keep a simple audit log for RAG or prompt use (prompt, sources, output, reviewer).
- Check an LLM answer against sources and record the check.
- Place LLM artifacts into the assurance case where they belong.

As safety-critical systems evolve, integrating emerging technologies presents both opportunities and challenges. The role of large language models (LLMs) in supporting development, analysis, validation, and maintenance of safety-critical systems is neither easy nor straightforward. Applying these technologies to safety-critical systems is particularly difficult and needs to be examined in a structured manner.

12.1 Key Terms and Applicable Standards

Before examining how these technologies affect safety-critical systems, it helps to establish clear definitions and identify the relevant regulatory landscape.

A **neural network (NN)** is a parametric function approximator trained on data. Unlike traditional software where behavior follows explicitly programmed rules, neural networks learn patterns from training examples and generalize to new inputs. In safety contexts, this means treating neural networks as components with defined operating boundaries, requiring test coverage that demonstrates acceptable behavior within those boundaries, and monitoring hooks that detect when inputs fall outside the validated domain.

A **large language model (LLM)** is a neural network trained on text, often refined with reinforcement learning from human feedback, that produces probabilistic text completions. These models can generate fluent and apparently knowledgeable text, but their outputs represent statistical patterns in training data rather than verified facts. For safety applications, the key principle is to treat LLM outputs as unverified until checked against trusted sources—they are starting points for human review, not authoritative answers.

Relevant standards and guidance continue to evolve as regulators grapple with these new technologies. Key frameworks include IEC 61508 (functional safety), ISO 26262/21448 (road vehicles), ISO/IEC 23894 (AI risk), the EU AI Act [7], NIST AI RMF [8], and UL 4600 [9]. Because this regulatory landscape is actively developing, practitioners should always confirm the latest edition of any standard and cite the exact part they apply.

These definitions are deliberately short. If you are new to AI safety, start by writing in plain language what you think the model does, where it will fail, and how you will notice. That short note often keeps the rest of the discussion grounded.

12.2 Self-driving Cars and Neural Networks: Defining the Limits of "Safety"

Case in Point. A perception model misclassifies reflective signage at dusk, causing erratic braking. Post-incident analysis shows the training set underrepresented low-sun-angle scenes. The fix adds targeted data collection and a brightness/contrast augmentation policy, plus a runtime plausibility check that gates braking on multi-sensor confirmation.

12.2.1 Defining Safety in Autonomous Systems

Safety-critical systems are characterized by the requirement to prevent or reduce harm through reliable and predictable operation. In the context of self-driving vehicles, this translates explicitly into collision avoidance, protection of passengers and pedestrians, and consistently predictable responses inverse and dynamic scenarios. See Figure 12.1 for a compact contrast between traditional safety engineering and data-driven autonomy.

Traditional safety approaches rely on deterministic behavior and well-defined operational parameters. However, autonomous systems introduce fundamentally different challenges:

- Non-deterministic decision-making based on probabilistic algorithms
- Open operational contexts with potentially infinite scenario variations
- Real-time adaptation to unforeseen conditions and situations
- Complex interactions with human-controlled vehicles and pedestrians

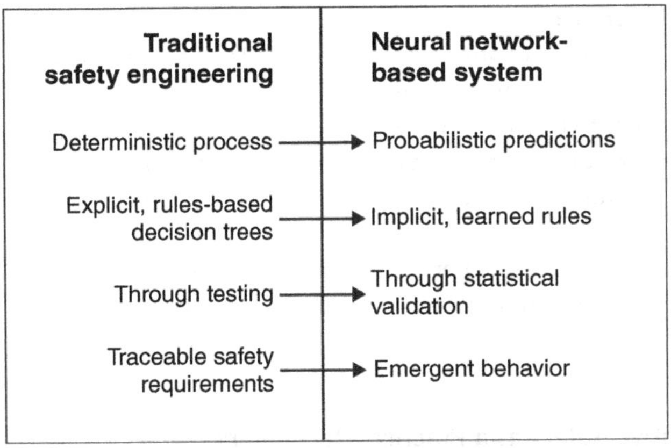

Figure 12.1 Conventional safety engineering vs. neural network-based systems.

12.2.2 Challenges of Neural Network-based Systems

Neural networks, despite their capabilities, present unique challenges:

- **Opacity (Black-box Problem):**
  - Neural network decision-making processes are inherently difficult to interpret, posing challenges for safety certification and accountability.
  - Traditional verification techniques that rely on understanding all possible execution paths become infeasible with multi-billion parameter networks.
- **Unpredictability and Bias:**
  - Biases present in training data and unforeseen edge cases can result in unpredictable or unsafe behaviors in real-world scenarios.
  - Small perturbations in inputs (e.g., modified or dirty road signs) can lead to dramatically different and potentially unsafe decisions.
- **Dynamic Environmental Complexity:**
  - Real-world environments for vehicles are unpredictable, continuously evolving, and extremely complex, conditions neural networks can find difficult to navigate reliably.
  - The infinite variability of potential scenarios makes exhaustive testing impossible.

Example: Edge Case Challenge in Neural Networks

```
Scenario: An autonomous vehicle approaches a partially
obscured pedestrian crossing during heavy rainfall at dusk.

Traditional Approach: Explicit rules for reduced speed in
poor visibility, heightened caution near crosswalks.
```

```
Neural Network Challenge:
- The combination of poor lighting, rain reflections, and
  partial obstruction creates a scenario potentially
  underrepresented in training data.
- The network might fail to recognize the pedestrian or
  misclassify the situation due to the unusual visual
  patterns.
- The decision-making process remains opaque, making it
  difficult to predict or diagnose failures in such edge
  cases.
```

12.2.3 Adopting Relative Safety as a Practical Benchmark

Given the above constraints, absolute safety—a state devoid of risk—may be practically unattainable. A more pragmatic approach is adopting a standard of relative safety, which compares autonomous system performance directly with human-operated counterparts. For example:

- If autonomous vehicles demonstrate measurable reductions in accident rates compared to human-driven vehicles, they can justifiably be considered safer despite the potential for occasional failures.
- This approach acknowledges that both human drivers and autonomous systems will experience failures, but focuses on overall harm reduction across the entire transportation system.
- Relative safety metrics might include:
  - Accidents per million miles driven
  - Severity of accidents when they do occur
  - Performance in high-risk scenarios (poor weather, complex intersections)
  - Response time to potential hazards

This viewpoint accepts inherent limitations and shifts focus towards meaningful, incremental improvements over existing safety levels.

12.2.4 Barriers to Safety Certification

Several critical barriers prevent current neural network systems from achieving traditional safety certification standards:

- **Explainability Limitations:**
  - The inability of neural networks to provide clear, deterministic explanations challenges traditional certification processes reliant on explicit logic.
  - Current standards require demonstrable evidence that all potential failure modes have been identified and mitigated—challenging with inherently opaque systems.

- **Continuous Learning Risks:**
  - Systems designed for dynamic learning in real-time risk unpredictable performance changes, undermining static safety assurances.
  - Changes in the model over time may invalidate previous safety analyses and certificates.
- **Existing Regulatory Constraints:**
  - Current safety standards such as ISO 26262 are primarily designed around deterministic, predictable systems, complicating their application to probabilistic neural network behaviors.
  - Regulatory frameworks struggle to define appropriate safety thresholds for systems whose behavior cannot be exhaustively tested.

12.2.5 Pathways Toward Safer Neural Network Implementations

Several promising strategies to improve neural network safety include:

- **Explainable AI (XAI):**
  - Developing neural network architectures capable of providing interpretable outputs and decision justifications.
  - Implementing attention mechanisms and feature visualization tools to highlight what aspects of inputs are driving decisions.
- **Hybrid Systems (Safety-by-design):**
  - Integrating deterministic, rule-based safety mechanisms alongside neural network components to ensure reliable fallback operations.
  - Implementing runtime monitoring systems that verify neural network outputs against safety constraints before execution.
- **Extensive Simulation and Validation:**
  - Using high-fidelity simulation environments to systematically expose neural network systems to diverse scenarios, improving resilience and reliability.
  - Implementing adversarial testing to deliberately challenge the neural network with difficult edge cases.
- **Formal Verification Approaches:**
  - Applying specialized formal verification techniques adapted for neural networks to provide mathematical guarantees about behavior boundaries.
  - Developing bounded verification methods that can establish safety properties for specific operational domains.

12.3 Using Large Language Models in Safety-critical Systems Engineering

For governance rhythms and evidence, see Section 5.4.

Case in Point. An LLM drafts a hazard checklist for a power subsystem. Reviewers find it fluent but incomplete: it omits ground-return faults. The team updates the prompt to include interface assumptions and requires a cross-check against a canonical failure-mode list. The LLM output becomes a *starting point*, not an authority, and gaps are caught in review.

LLMs, such as GPT and similar technologies, can change safety-critical system engineering by automating and improving key activities. Figure 12.2 summarizes where LLMs can assist (requirements, hazard analysis, evidence hygiene) and where they must not replace engineering judgment.

12.3.1 Requirements Analysis

LLMs can streamline requirements management:

- **Natural Language Processing (NLP):**
  - Automatically parsing, clarifying, and ensuring precision in requirements documentation.
  - Identifying ambiguities, inconsistencies, and implicit assumptions in natural language requirements.

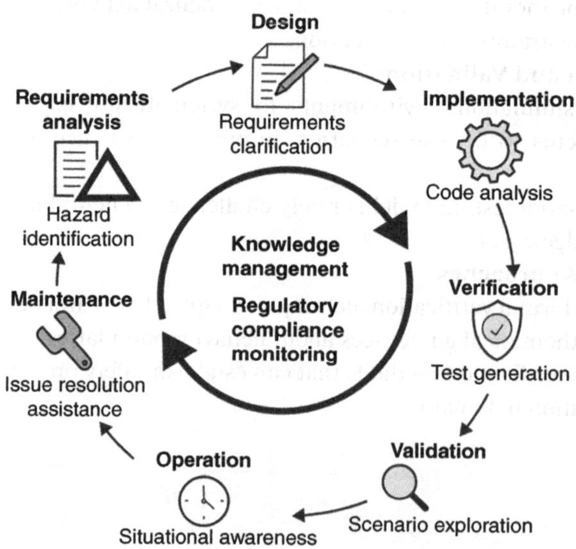

Figure 12.2 Roles of large language models in the safety-critical systems lifecycle.

- **Traceability Enhancement:**
  - Automatically mapping and documenting requirements relationships, improving traceability throughout system lifecycle stages.
  - Generating traceability matrices that link requirements to verification methods, design components, and regulatory standards.

Example: LLM-enhanced Requirements Analysis

```
Original Requirement: "The system shall respond in a timely
 manner to emergency situations."

LLM Analysis:
- Ambiguity detected: "timely manner" is subjective and not
    measurable
- Missing definition: "emergency situations" not
    specifically defined
- Related standards: IEC 61508 requires quantifiable
    response requirements

LLM-Suggested Refinement:
"The system shall detect and initiate response to defined
    emergency conditions within 200ms of occurrence, and
    complete all required safety actions within 2 seconds.
    Emergency conditions include: [list of specific
    conditions]."

Additional suggestions:
- Add verification method: timing analysis and fault
    injection testing
- Link to related requirements: fault detection, alarm
    system requirements
```

12.3.2 Hazard Identification and Scenario Generation

LLMs can automate thorough hazard analyses:

- **Knowledge Integration:**
  - Synthesizing historical safety data, standards, and best practices to proactively suggest potential hazards.
  - Drawing connections between seemingly unrelated system aspects to identify emergent risks.

- **Scenario Creation:**
  - Generating realistic and challenging failure scenarios, improving robustness assessments.
  - Creating "what-if" analyses that human analysts might overlook due to cognitive biases or domain-specific blinders.

12.3.3 Improved Verification and Validation

LLMs can improve verification and validation processes:

- **Automated Test Case Generation:**
  - Producing thorough test cases covering edge conditions and unusual scenarios often overlooked by manual processes.
  - Generating test cases directly from requirements specifications with appropriate coverage metrics.
- **Code and Logic Analysis:**
  - Identifying non-compliance, coding errors, or logical inconsistencies in system implementations through automated static analysis.
  - Comparing implementation code and drawings against requirements to verify completeness and correctness.

12.3.4 Risk Assessment and Analysis

LLMs offer support for systematic risk analyses:

- **Structured Methodology Support:**
  - Guiding engineers through methodologies like failure modes, effects, and criticality analysis (FMECA) and fault tree analysis (FTA), highlighting potential risks and mitigation measures.
  - Ensuring completeness and consistency in risk analysis documents.
- **Comparative Historical Analysis:**
  - Comparing current projects to past implementations to uncover common risk factors and proven mitigation strategies.
  - Using lessons learned from similar systems or incidents to improve current designs.

12.3.5 Documentation, Training, and Communication

LLMs can transform documentation practices and training methods:

- **Dynamic Documentation Generation:**
  - Automatically updating documentation to reflect real-time changes in designs, standards, or compliance requirements.

  - Generating consistent, well-structured documentation across different system components.
- **Interactive and Adaptive Training Tools:**
  - Offering conversational interfaces for interactive, context-aware training and knowledge reinforcement.
  - Creating personalized learning pathways based on individual roles and expertise levels.

12.4 Challenges and Considerations in Using LLMs

While the potential of LLMs is real, challenges remain.

12.4.1 Technical Limitations

- **Accuracy and Reliability Concerns:**
  - Ensuring consistently accurate and reliable outputs from LLMs, necessitating ongoing human oversight.
  - "Hallucination" phenomena where LLMs generate plausible but factually incorrect information present risks in safety contexts.
- **Domain-Specific Context Limitations:**
  - Addressing limitations in understanding nuanced, domain-specific technical and operational contexts.
  - Ensuring LLMs have access to current, relevant standards and best practices.
- **Validation Needs:**
  - Implementing verification procedures to confirm LLM outputs meet safety and compliance requirements.
  - Developing appropriate testing methodologies for LLM-based tools.

12.4.2 Governance Frameworks

Several key governance considerations must be addressed:

- **Human Oversight:**
  - Establishing clear protocols for human verification of critical LLM outputs.
  - Defining appropriate trust levels for different applications of LLM technology.
- **Accountability and Responsibility:**
  - Determining accountability structures when LLMs contribute to system design or verification.
  - Addressing legal and ethical implications of automated decision-making in safety contexts.

- **Change Management:**
  - Developing protocols for managing LLM version changes and training data updates.
  - Ensuring traceability between specific LLM versions and safety-critical outputs.

12.5 Case-specific Considerations

Different safety-critical domains present unique challenges and opportunities when incorporating advanced technologies. While the fundamental principles of safety assurance apply across domains, the specific regulatory frameworks, operational contexts, and risk profiles vary considerably.

12.5.1 Medical Applications

Medical applications of AI face the dual challenge of demonstrating clinical benefit while ensuring patient safety. Neural networks used for diagnostic support require clinical validation protocols that go beyond software verification—the model must be shown to improve patient outcomes, not just classify images accurately. This requires clinical trials with appropriate patient populations and endpoints, conducted under regulatory oversight.

LLMs offer the potential to improve medical documentation quality and regulatory compliance, areas where the volume and complexity of requirements often overwhelm human capacity. However, patient privacy considerations introduce additional constraints on training data and model deployment. Medical AI systems must comply with privacy regulations while still having access to sufficient data to train and validate—a tension that requires careful architectural and governance solutions.

12.5.2 Aerospace and Aviation

Aviation has some of the most mature and stringent safety certification frameworks, presenting both challenges and opportunities for AI integration. Certification challenges are particularly demanding under regulations like DO-178C, which were designed for deterministic software and do not map neatly onto probabilistic neural network behavior. Adapting these frameworks or developing appropriate alternatives remains an active area of regulatory development.

Hybrid decision systems with human pilots in the loop present specific human factors considerations. The interaction between automated recommendations and human judgment must be carefully designed to avoid automation complacency while still providing the benefits of AI assistance. LLMs can improve maintenance documentation and safety reporting systems, areas where the volume of data often overwhelms human processing capacity, improving overall fleet safety without introducing AI into flight-critical decision loops.

12.5.3 Industrial Control Systems

Industrial control systems often operate in environments where legacy system integration presents unique challenges for new AI technologies. Many industrial installations have decades-old control systems that were never designed to interface with modern AI components. Integration must carefully preserve the safety properties of existing systems while adding new capabilities.

Physical safety interlocks and independent protection layers remain necessary alongside machine learning systems—AI should augment, not replace, proven safety mechanisms. The fundamental principle of defense in depth applies: no single technology, including AI, should be the sole barrier against catastrophic outcomes. LLMs can improve alarm management and operator support without compromising core safety functions, helping operators process the flood of information that modern plants generate while keeping safety-critical decisions in the hands of proven systems and trained personnel.

12.6 Emerging Standards and Best Practices

The landscape of standards for emerging technologies in safety contexts is rapidly evolving: Figure 12.3 outlines governance and assurance patterns that are converging across industries.

- **Current Developments:**
  - ISO/IEC TR 24028 (AI trustworthiness assessment)
  - IEEE 7000 series for ethical AI
  - Domain-specific guidance from regulatory bodies
- **Industry Adaptations:**
  - Developing consensus around acceptable evidence for neural network assurance
  - Defining appropriate safety margins and operational limitations
  - Creating standardized simulation and verification environments

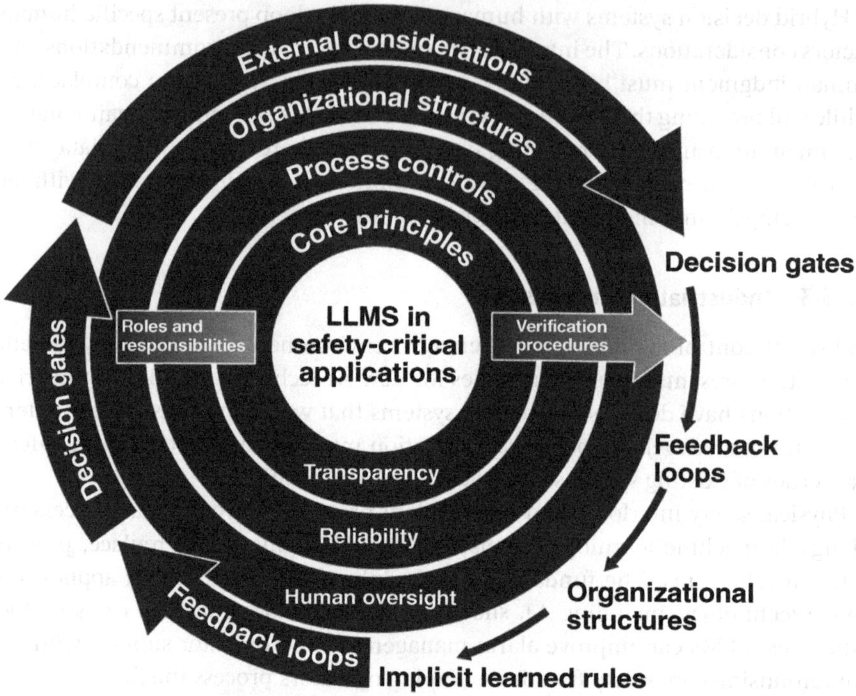

Figure 12.3 Governance framework for LLM implementation in safety-critical systems.

12.7 Concluding Reflections

The integration of emerging technologies such as neural networks and LLMs into safety-critical systems shows the need for thoughtful, well-structured approaches. While absolute safety may remain an elusive goal in certain autonomous applications, the practical adoption of relative safety benchmarks provides a realistic framework for progress. LLMs offer real advantages in system analysis, documentation, and risk management, but their use must be carefully managed, validated, and continuously supervised.

Three key principles should guide the adoption of new technologies in safety-critical contexts:

1. Maintain appropriate human oversight with clearly defined responsibilities
2. Implement layered safety approaches that do not rely exclusively on emerging technologies
3. Establish validation approaches specific to the unique challenges of these technologies

By addressing these complexities, organizations can use new technologies to improve safety, delivering safer systems.

Key Takeaways

- LLMs add risk if unchecked but help when governed—retain auditability and keep assessor-ready artifacts.
- Prefer design-time assistance (review, traceability); any run-time autonomy that can influence hazards must meet full functional-safety controls.
- Keep a human owner and explicit acceptance criteria; maintain auditable evidence and treat certain uses as high-risk under applicable regulations.
- RAG/agent patterns need guardrails—curate/version the corpus, log prompts/sources/outputs, require sources; agent loops should *propose* then *wait* (no unattended changes).
- Use relative-safety benchmarks where absolute safety is impractical—compare to human baselines and focus on overall harm reduction.
- Recognize certification barriers for neural nets (explainability limits, drift, deterministic-standards mismatch) and mitigate with XAI, hybrid safety-by-design, and simulation/adversarial testing.
- Anchor to evolving standards and layered safety (e.g., ISO/IEC TR 24028, IEEE 7000); apply human oversight and independent protection layers.

13

Final Considerations

After reading this chapter, you should be able to:

- Summarize the few practices from this book you will carry into your next project.
- Build a short personal checklist (one page) you will actually use.
- Pick one improvement to your team's current safety process and plan the first step.
- Note where your evidence is weakest and how you will raise it on the next milestone.

Safety-critical systems demand engineering rigor, technical precision, and ethical responsibility. Throughout this book, we have covered the principles, processes, and challenges involved in designing, developing, and operating these systems. As we conclude, it helps to reflect on the lessons learned and outline paths for further exploration and application.

13.1 Key Takeaways Revisited

13.1.1 Safety as an Ongoing Commitment

Safety is not simply a static achievement or state but a continuous, dynamic process requiring ongoing review and vigilance. Maintaining safety requires:

- **Proactive Design and Engineering:**
  - Identifying and mitigating potential hazards early and consistently throughout the lifecycle.
  - Applying safety-by-design principles from initial concept through implementation.

- **Continuous Analysis and Feedback Loops:**
  - Regularly revisiting assumptions, adjusting strategies, and integrating new data and experiences.
  - Implementing formal mechanisms for capturing and incorporating lessons learned.
- **Monitoring and Adaptation:**
  - Establishing operational monitoring mechanisms to detect and address risks promptly in real-world settings.
  - Creating responsive systems capable of adapting to changing conditions and emerging threats.

Example: Implementation of Continuous Safety Improvement

```
A medical device manufacturer implements a "Safety
Improvement Cycle" with these components:

1. Monthly safety performance metrics review with
   cross-functional teams
2. Quarterly hazard analysis updates integrating field
   feedback
3. Semi-annual formal review of all safety assumptions
   with verification
4. Annual independent assessment of safety mechanisms
5. Continuous monitoring system that automatically flags
   deviations from expected operational parameters

This structured approach ensures safety remains an ongoing
focus rather than a one-time certification activity, leading
to continuous refinement of critical safety mechanisms.
```

13.1.2 Navigating Trade-offs

Designing safe systems involves balancing multiple, often conflicting, objectives:

- **Functionality vs. Safety:**
  - Delivering intended functionality without compromising system safety and user protection.
  - Recognizing when adding features introduces unnecessary complexity and unacceptable safety risks.
- **Cost vs. Reliability:**
  - Achieving good safety within realistic budget constraints, balancing economic viability with system reliability.

  - Identifying where additional investment yields safety improvements versus diminishing returns.
- **Speed vs. Thoroughness:**
  - Ensuring timely project completion without neglecting necessary rigor and thorough safety evaluations.
  - Developing strategies to accelerate safety processes without compromising their effectiveness.

Effective decision-making under these conditions requires clarity of purpose, transparency in reasoning, and ethical commitment.

13.1.3 Importance of Independent Perspectives

Independent assessment, verification, and validation matter for maintaining system safety. Independence helps:

- **Avoid Cognitive Biases and Groupthink:**
  - Overcome confirmation bias, where teams unconsciously filter information to support existing beliefs.
  - Challenge normalized deviance, where gradual shifts from safe practices become accepted.
- **Challenge Prevailing Assumptions:**
  - Question design decisions that internal teams may take for granted.
  - Identify blind spots in safety analysis that insiders might overlook.
- **Incorporate Diverse Perspectives:**
  - Enrich analyses through varied backgrounds and experience levels.
  - Strengthen safety integrity through multiple independent verification paths.

13.1.4 Emerging Technologies: Opportunities and Risks

Technologies such as artificial intelligence (AI), neural networks, large language models (LLMs), the Internet of Things (IoT), and advanced robotics offer advancements but simultaneously introduce complexity and new risk dimensions. Successfully integrating these technologies demands:

- **Developing New Methodologies:**
  - Creating verification approaches specifically tailored to non-deterministic AI systems.
  - Establishing frameworks for safety assurance in highly connected IoT ecosystems.

- **Adapting Regulatory Frameworks:**
  - Evolving standards to address novel behaviors and failure modes of autonomous systems.
  - Developing principles-based approaches where prescriptive standards cannot keep pace with innovation.
- **Testing Thoroughly:**
  - Implementing simulation environments to test AI systems across diverse scenarios.
  - Developing requirements that allow safety assessment of complex algorithms.

13.1.5 Human-centric Safety Approaches

At the heart of every safety-critical system are people who design, operate, and maintain them. Effective safety management requires:

- **Building a Strong Safety Culture:**
  - Building organizational values that prioritize openness, accountability, and continuous learning.
  - Establishing mechanisms that encourage reporting of concerns without fear of reprisal.
- **Providing Training:**
  - Developing in-depth training programs tailored to specific roles and responsibilities.
  - Regularly refreshing safety knowledge to prevent complacency and skill deterioration.
- **Addressing Human Factors Proactively:**
  - Designing interfaces and workflows with careful consideration of human limitations.
  - Implementing error-resistant designs that anticipate and mitigate potential human errors.

13.2 Exploring Further: Practical Steps and Recommendations

Anchor your plan to three pillars: independence (Section 2.1), V&V (Section 5.4), and monitoring (Section 11.1).

Case in point (first 90 days). A team selects three moves: (1) introduce a weekly risk sync with a shared assumption ledger; (2) add boundary-value and fault-injection test items to two modules; (3) pilot an independence checklist for one review. Within one quarter, two latent assumptions are closed, a boundary defect is caught pre-integration, and review friction drops as roles become explicit.

13.2.1 Tailoring Safety Practices to Your Context

While the principles discussed in this book are universally applicable, they must be customized to your organizational context and operational domain:

- **Conduct Thorough Assessments:**
  - Evaluate existing organizational processes and safety practices against industry standards.
  - Use structured methods like maturity models to benchmark current capabilities.
- **Identify Strategic Gaps:**
  - Analyze discrepancies between your current procedures and best practices.
  - Create a prioritized gap analysis showing areas needing immediate attention.
- **Prioritize Improvements:**
  - Focus initial efforts on addressing gaps that provide the greatest immediate and long-term safety improvements.
  - Develop phased implementation plans that balance ambitious goals with practical constraints.

13.2.2 Staying Ahead of Technological Evolution

Remaining conversant with emerging technologies and methodologies is important:

- **Evaluate New Technologies:**
  - Regularly assess emerging tools and methodologies for their safety implications and applications.
  - Establish pilot programs to explore promising technologies in controlled environments.
- **Stay Informed About Advances:**
  - Follow developments in explainability, validation techniques, and evolving safety standards.
  - Participate in professional communities focusing on safety-critical systems.
- **Invest in Ongoing Education:**
  - Create structured learning pathways for safety professionals at all organizational levels.
  - Allocate resources for training in emerging safety methodologies and technologies.

13.2.3 Contributing to and Shaping Industry Standards

Active participation in standardization processes benefits your organization and the broader industry:

- **Engage with Professional Bodies:**
  - Join relevant standards committees and working groups in your industry sector.
  - Share insights, challenges, and practical experiences to inform standards development.
- **Collaborate with Industry Peers:**
  - Establish cross-organizational forums to discuss common safety challenges.
  - Develop pre-competitive collaboration on foundational safety approaches.
- **Document and Share Case Studies:**
  - Publish anonymized lessons learned to benefit the wider safety community.
  - Contribute to the collective knowledge based on effective safety practices.

13.2.4 Building Organizational Resilience

Resilient organizations are better equipped to manage challenges and maintain safety integrity:

- **Implement Response Protocols:**
  - Develop incident response and crisis management frameworks.
  - Regularly test these protocols through realistic simulation exercises.
- **Encourage Organizational Learning:**
  - Create systems for capturing, analyzing, and disseminating lessons from near-misses and incidents.
  - Implement formal processes for integrating insights into improved practices.
- **Develop Adaptive Capacity:**
  - Build flexibility into safety processes to handle unexpected situations.
  - Train teams to recognize when standard procedures may need adaptation for novel conditions.

13.2.5 Structured Professional Development

Safety engineering excellence requires deliberate skill development:

- **Identify Core Competencies:**
  - Map the technical, analytical, and communication skills needed for safety roles.
  - Assess current capabilities against this competency framework.

- **Create Personal Development Plans:**
  - Establish specific growth objectives aligned with organizational and industry needs.
  - Include formal education, mentoring, and practical experience components.
- **Pursue Recognized Certifications:**
  - Obtain relevant certifications that validate specialized safety knowledge.
  - Maintain credentials through continuous professional development.

13.3 Exploring New Domains and Cross-disciplinary Applications

The principles and practices of safety-critical systems have applications in adjacent fields and innovative domains:

- **Cybersecurity Integration:**
  - Apply safety-critical methodologies to improve security risk assessment and management.
  - Develop integrated approaches addressing the safety-security interface in connected systems.
- **Environmental Sustainability:**
  - Adapt safety risk frameworks to evaluate and mitigate environmental impacts.
  - Implement safety-oriented design principles in environmentally sensitive applications.
- **Smart Infrastructure:**
  - Apply safety-critical thinking to the development of interconnected urban systems.
  - Integrate resilience considerations into critical infrastructure planning.
- **Public Health Systems:**
  - Transfer safety-critical monitoring and response approaches to public health surveillance.
  - Implement systematic hazard analysis methods for health emergency preparedness.

13.4 Looking to the Future: Your Role as a Safety Leader

The future of safety-critical systems promises increasing complexity, interconnectivity, and autonomy. As professionals, you will shape this future:

- **Embrace Continuous Learning:**
  - Maintain intellectual curiosity and openness to new approaches and technologies.
  - Develop the habit of questioning established practices as contexts evolve.
- **Build Ethical Frameworks:**
  - Develop ethical decision-making models for safety trade-offs.
  - Advocate for transparent consideration of societal impacts in safety decisions.
- **Champion Responsible Innovation:**
  - Balance technological advancement with safety assurance.
  - Promote safety as an enabler rather than an impediment to innovation.
- **Develop Future Safety Talent:**
  - Mentor emerging safety professionals to build the next generation of expertise.
  - Contribute to educational programs that prepare students for safety challenges.

Key Takeaways

- This book provides a guide with insight, methodologies, and reflective practices for the engineering of safety-critical systems.
- The journey ahead will be shaped by your actions, your willingness to tackle complex problems, and your dedication to continuous improvement.
- Effective safety professionals must balance technical expertise with organizational awareness, ethical judgment, and adaptability.
- They must navigate between established best practices and emerging approaches, between regulatory compliance and innovation, between detailed analysis and practical implementation.
- Safety is not a destination—it is a continual journey of vigilance, innovation, and responsibility.

References

1 IEC. Functional safety of electrical/electronic/programmable electronic safety-related systems (parts 1–7), 2010. 2nd edition.
2 MISRA. MISRA C:2012 guidelines for the use of the C language in critical systems, 2012.
3 U.S. Department of Defense. MIL-STD-882E: Department of Defense Standard Practice: System Safety, 2012.
4 European Space Agency. Flight 501 failure: Inquiry board report, 1996. Public inquiry report on Ariane 5 Flight 501 failure. https://www.esa.int/Science_Exploration/Space_Science/Ariane_501_failure.
5 Nancy G. Leveson. *Engineering a Safer World: Systems Thinking Applied to Safety.* MIT Press, Cambridge, MA, 2012. ISBN 978-0262016629.
6 Joint Authorities Technical Review. Boeing 737 MAX flight control system joint authorities technical review, 2019. FAA-commissioned report on MCAS and certification. https://www.faa.gov.
7 European Union. Artificial intelligence act, 2024. OJ L, 2024.
8 National Institute of Standards and Technology. AI risk management framework 1.0, 2023.
9 *UL 4600: Standard for Safety for the Evaluation of Autonomous Products.* UL Standards & Engagement, 2nd edition, 2022.

Further Readings

10 CENELEC. Railway applications – communication, signalling and processing systems – safety-related communication in transmission systems, 2010.
11 CENELEC. Railway applications – communication, signalling and processing systems – software for railway control and protection systems, 2011.
12 CENELEC. Railway applications – the specification and demonstration of reliability, availability, maintainability and safety (RAMS) – part 1: Generic RAMS process, 2017.

13 CENELEC. Railway applications – the specification and demonstration of RAMS – part 2: System approach to safety, 2017.

14 CENELEC. Railway applications – communication, signalling and processing systems – safety-related electronic systems for signalling, 2018.

15 Clifton A. Ericson II. *System Safety Engineering and Risk Assessment: A Practical Approach*. Wiley, Hoboken, NJ, 2nd edition, 2015. ISBN 978-1118940389.

16 David J. Smith and Kenneth G. L. Simpson. *The Safety Critical Systems Handbook*. Elsevier, Oxford, UK, 5th edition, 2018. ISBN 978-0128051216.

17 IEC. Fault tree analysis (FTA), 2006.

18 IEC. Medical device software – software life cycle processes, 2015.

19 IEC. Failure modes and effects analysis (FMEA and FMECA), 2018.

20 ISO/IEC. Information technology – artificial intelligence – risk management, 2023.

21 SAE International. Guidelines for development of civil aircraft and systems, 2023.

22 SAE International. Guidelines for conducting the safety assessment process on civil airborne systems and equipment, 2023.

23 Sneha S. Shetiya, Divya Garikapati, and Veeraja Sohoni. FTA generation using GenAI with an autonomy sensor usecase, 2024. arXiv:2411.15007v1.

Index

b

e

i

p

u

Printed in the USA/Agawam, MA
August 28, 2026

909771.002